INTRODUCTION TO FLUID KINEMATIC

ANALYTIC STUDY OF FLUID MECHANIC, FLOW DYNAMISM, STRESS, STRAIN, AND VORTICITY

PAPRI BHATTACHARJEE

Copyright © 2024 PAPRI BHATTACHARJEE

All rights reserved.

ISBN:9798324332792

CONTENTS

1. KINEMATICS OF FLUIDS

INTRODUCTION
TYPES OF FLOWS
METHODS OF DESCRIBING FLUID MOTION
MATERIAL, LOCAL AND CONVECTIVE DERIVATIVE
FLOW VISUALIZATION
VELOCITY POTENTIAL FUNCTION
EQUIPOTENTIALS
STREAM FUNCTION
BOOSTER CAPSULE : MACH NUMBER
HIGHER ORDER THINKING SKILL QUESTION

2. DYNAMICS OF FLOWS

INTRODUCTION
UNIFORM FLOW
SOURCE FLOW
SINK FLOW
FREE-VORTEX FLOW
PRINCIPLE OF CONSERVATION OF MASS
EQUATION OF CONTINUITY
DERIVATION USING THE DIVERGENCE THEOREM
EQUATION OF CONTINUITY FOR AN INFINITESIMAL CONTROL VOLUME
EQUATION OF CONTINUITY IN CARTESIAN COORDINATES
EQUATION OF CONTINUITY IN CYLINDRICAL COORDINATES
EQUATION OF CONTINUITY IN POLAR COORDINATES
DEDUCTIONS
BOOSTER CAPSULE: NO SLIP CONDITION
HIGHER ORDER THINKING SKILL QUESTION

3. STRESS

INTRODUCTION
CAUCHY'S STRESS POSTULATES
TYPES OF STRESS
EXPRESSION FOR STRESS TENSOR
SYMMETRY OF STRESS TENSOR
PRINCIPAL STRESSES

TRANSFORMATION OF STRESS COMPONENTS

MOHR'S CIRCLE

BOOSTER CAPSULE : DRAG AND LIFT

HIGHER ORDER THINKING SKILL QUESTION

1. Calculate the viscous stress induced in Couette Flow.
2. Calculate the viscous stress induced in Haigen Flow

4. STRAIN

INTRODUCTION

GENERAL ANALYSIS OF FLUID MOTION

EXPRESSION FOR STRAIN TENSOR

TRANSFORMATION OF STRAIN TENSORS

RELATION BETWEEN STRESS AND STRAIN TENSORS

NAVIER-STOKES EQUATION

BOOSTER CAPSULE : TORQUE

HIGHER ORDER THINKING SKILL QUESTION

5. VORTICITY

INTRODUCTION

CONSTITUTIVE EQUATIONS

VORTEX FLOW

KELVIN CIRCULATION THEOREM

IRROTATIONAL FLOW

BOOSTER CAPSULE

HIGHER ORDER THINKING SKILL QUESTION

ACKNOWLEDGMENTS

I would like to acknowledge with appreciation for the valuable comments, suggestions, constructive criticisms from the following evaluators and reviewers:

Prof. Yunus A. Cengel

University of Nevada, Reno

Prof. John M. Cimbala

Pennsylvania State University

Prof. R.K. Bansal

Delhi College of Engineering, India

Prof. M.D. Raisinghania

S.D. College, Uttar Pradesh, India

Prof. S.K. Sharma

Himachal Pradesh University, India

Prof. Arup Kumar Das

IIT Roorkee, India

Prof. Suman Chakraborty

IIT Kharagpur, India

Prof. Sreenivas Jayanti

IIT Madras, India

Prof. Niket Kaisare

Programmer, IIT Madras, India

Prof. Vijayender

Educator, Founder of YP channel, India

I am thankful to Gauhati University for their valuable

contributions, particularly suggestions for the improvisation of contents of this book, and numerical problems, and their critical review of the entire manuscript. I also thank to the Amazon KDP University for contributing several end-of-chapter problems, and for reviewing the book and pointing out numerous errors.

Lastly, special thanks should go to my parents for their continued patience, understanding, and support throughout the preparation of this book, which involved many long hours when they had to handle family concerns on their own as their daughter's face was stick to the laptop screen.

The author also thankful to the publisher for critically reviewing the manuscript.

PAPRI BHATTACHARJEE

AUTHOR'S NOTE

Fluid Mechanic is a vast subject deals with study of dynamism of the fluid in different conditions with correlated with parameters such as pressure, velocity, vorticity, torque etc. All of us, aware of the wide application of fluid mechanics specially in rocket engines, wind turbines, oil pipelines, and air conditioning systems, groundwater management and controlling devices, etc. So, working principles of various technical devices are based on the basic parameter such as the pressure, velocity, vorticity, torque, etc which are study under Fluid Kinematics. Therefore, we must have the basic knowledge of fluid kinematic. As the total, here the book titled **"INTRODUCTION TO FLUID KINEMATIC** try to approach the basic understanding of the fundamentals of fluid kinematic. Here, introduction of Chapter 1 brief about the basic knowledge of fluid. Then Chapters deals with Equation of continuity and the Equation of motion in inviscid fluids with their derivation. This book is not free of limitations. Though author tried her best to make the book free of errors and limitations, I apologize if any error is found. Please feel free to contact author and reflect your comments to improve this book in next editions.

THANK YOU

KEEP READING !

KEEP LEARNING !

Papri Bhattacharjee
Email id: papri201514@gmail.com

1. KINEMATICS OF FLUIDS

INTRODUCTION

Life on earth is possible due to the presence of atmosphere (mixture of gases) and water (liquid). Also, majority portion of earth's surface is covered with water (liquid). In the same way, liquids and gases are very important matter whose applications are discuss as follow:

1. Human heart is filled with blood which is one type of liquid that circulate throughout the entire body and pumping at 72 beats/minute. While, we can observe the heart beat rate increases when pressure in blood circulation is high.
2. Ocean is good example of salt water, again consider as liquid that have wide variety of implications for the existence of life on earth.
3. Refrigerator works on the principle of condensation of liquid and conduction of gases within the design of the device.
4. Ships are moves on water (consider as a liquid).
5. Flow of liquids through pipes are observed in both commercial as well as domestic establishments.
6. Human breathe in and out CO_2 and O_2 which are consider as gas.
7. Plants are rely on water, air, and sunlight for their survival.
8. 70% of human body is filled with water which is consider as a liquid.
9. Birds are flying in air, which is nothing but a combination of various gases.

10. Engine oil which is consider as a liquid provides fuel for the functioning of automobile.
11. Hydraulic device are those device where liquids are being used as a fuel for their functionality. E.g., hydraulic lift, hydraulic brakes in car, hydraulic press, etc.
12. Gases and liquid are being used for the working of coolants such as AC, cooler, refrigerator, etc.
13. We have observed various gases are being used for launching of rockets, missiles, aircrafts.
14. Wind turbine use wind (gas) for the generation of energy.
15. Similarly, water turbines are implanted in dam which energize the flow of water (liquid) for producing electricity.
16. Many more examples are there

Thus, we can say that a substance in the liquid or gas form is consider as a **fluid.** In other words, A fluid is a substance which is continuously deforms under the action of applied force, irrespective of its intensity. A fluid includes both liquids and gases as shown in figure.

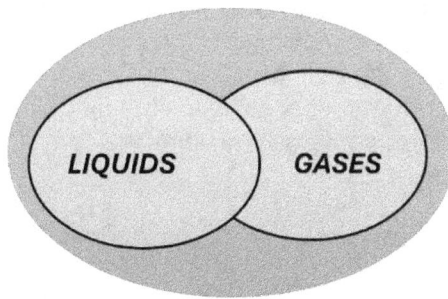

The difference between the solid and fluid is in terms of their ability to resist applied force which tends to change its shape and dimension. A solid can resist applied force by deforming whereas a fluid deforms continuously as it can resist applied force partially. Thus, at some angle, solid stop deforming its dimension while a fluid never stops deforming and makes rate of strain under the influence of applied force. The distinction between a solid and a fluid will be clearer from the figure.

INTRODUCTION TO FLUID KINEMATICS

$$F = \text{rate of change of momentum}$$

$$\Rightarrow F = \frac{d}{dt}(mv)$$

SOLIDS

Mass is constant.
i.e., $m = constant$

$$\Rightarrow F = m\frac{dv}{dt} = ma$$

$Force = mass \times acceleration$

FLUIDS

$$\Rightarrow F = \frac{dm}{dt}dv = \dot{m}dv$$

$Force = mass\ flow\ rate \times change\ in\ velocity$

Any fluid is consists of large number of molecules and the properties of the fluid is depends on the behaviour of these molecules. Let us consider, the pressure of a gas in a closed container is mainly due to the intermolecular collision between the molecules and the walls of the container. Moreover, one does not need to know the behaviour of individual molecule to determine the pressure. Thus, if we attach pressure gauge to the container, we can determine the pressure. So, we can say that the study of individual molecule is not appropriate way in concerned with the study of the motion of fluids or system in contact with fluids. Thus, we consider the macroscopic behaviour of a fluid from the point of view of numerical calculation. *Continuum Hypothesis* allows to study the fluids with the assumption that a fluid is continuously distributed in a given space and analysis involves the bulk behaviour of molecules of the fluid. This continuum theory explains the fluid particle defined as the fluid contained within the physically infinitesimal volume.

In general, we define density or velocity of a fluid is at a particular point. Moreover, we say that "fluid motion at a point is certain meters per second" or "density at a point is certain kg per square meter." These definitions are based on average movement of molecules moves through a small volume approaches to any particular point. This calculation will be suitable only when we consider small volume of the fluid. In addition, given volume should be sufficient to hold enough molecules to prevent the average values of various properties. Also, it is comparatively small in relation to the fluid system so

INTRODUCTION TO FLUID KINEMATICS

that it can consider as a point in the fluid system. Thus, measurable properties such as density, pressure, temperature, and bulk velocity are assumed to be well-defined at "infinitesimal" volume elements, which are small in comparison to the system's characteristic length scale. As velocity, density, pressure, and temperature vary from point to point, thus these are consider as the continuum of continuous media. Also, according to continuum theory, we analysis the macroscopic behaviour of fluid results into fluid is referred to as a continuum of matter.

Applicability of continuum theory is restricted to certain fluid flows such as supersonic flow and nanofluids. There are numerical methods are used to study those fluid flow where the continuum hypothesis fails. The Knudsen number, defined as the ratio of the molecular mean free path to the characteristic length scale. The Knudsen number defined as

$$Kn = \frac{\lambda}{L}$$

Where λ = molecular mean free path

L = characteristic length scale

The Knudsen number is used to check the validity of continuum hypothesis on any particular fluid. The following classification is helpful to choose appropriate mechanics model of a fluid flow:

- ❖ $Kn < 0.1 \Rightarrow$ fluid follow continuum hypothesis
- ❖ $0.01 < Kn < 0.1 \Rightarrow$ fluid flow with slip condition
- ❖ $0.1 < Kn < 10 \Rightarrow$ Transitional fluid flow
- ❖ $Kn > 10 \Rightarrow$ free molecular fluid flow (Knudsen Flow)

The study of fluid flow with high Knudsen number includes the movement of microfluidics, fluids used in MEMS devices, movement of fluids through the exosphere.

TYPES OF FLOWS

There is a wide variety of fluid flow problems found while studying fluid mechanics. In generalization, fluid flows can be grouped based on their characteristics. Some of the classifications are mentioned below:

COMPRESSIBLE & INCOMPRESSIBLE FLOW

We observed that the volumes (or density) of real fluids either expanded or compressed under the action of pressure or temperature. Thus, compressibility is the parameter of fractional change of volume (or density) of a fluid. So, the fluids whose volume (or density) varied with pressure or temperature is known as compressible fluid. On other side, the fluid whose volume (or density) does not least affected by the temperature or pressure is considered as incompressible fluid. In other words, the fluids whose volume (or density) remains constant comes under the category of incompressible fluid. A rigorous incompressible real fluid does not exist. Moreover, liquids are generally least compressible by pressure or temperature, such as water , the liquids flow referred as an incompressible fluid flow. Also, an ideal fluid taken as an incompressible fluid for the theoretical study purpose. Gas flows are compressible to a large extend but can be considered as incompressible if the volume (or density) variation are close to 5 %. The launching of rockets, spacecraft, and similar system involves high speed gas flows, therefore these gas flows can expressed in terms of **Mach number.**

It is observed that the volume (or density) of a fluid vary with the fluctuation in pressure and temperature. Usually, a fluid expands when pressure is released while the fluid's volume (or density) contracts when pressurized. On the same way, functioning of hot air balloons using natural convection depends on the temperature of the atmosphere. Moreover, the amount of volume (or density) vary differently for different fluids.

COMPRESSION BY PRESSURE

Let us consider a closed entropy, where it has been

INTRODUCTION TO FLUID KINEMATICS

observed that entropy contracts under the action of pressure force on the fluid, while it will expands on releasing the pressure from the entropy. So, we can say that fluids like solid can compress or de-compress under the action of pressure force. In fluid dynamics, compressibility of a fluid define in terms of bulk modulus of elasticity (or **coefficient of compressibility**) , can be expressed as

$$k = -V\left(\frac{\Delta P}{\Delta V}\right)_T = \rho\left(\frac{\Delta P}{\Delta \rho}\right)_T$$

Note that bulk modulus of elasticity of an incompressible fluid flow is ∞ as $\Delta V = 0$

Note that the specific volume and pressure are inversely proportional which indicates that pressure decrease on increasing volume (or density) and thus $\frac{\Delta P}{\Delta V}$ is a negative quantity so that **bulk modulus** k is a positive quantity.

Now recall the definition of specific volume given by

$$\rho = \frac{1}{V}$$

$$\Rightarrow d\rho = -\frac{dV}{V^2}$$

$$\Rightarrow \frac{d\rho}{\rho} = -\frac{1}{1/V}\left(\frac{dV}{V^2}\right) = -\frac{dV}{V}$$

Thus, the differential changes in the density and specific volume of a fluid are equal in magnitude but opposite in sign. Let consider the case of ideal gas where equation of state holds as

$$P = \rho RT \Rightarrow \frac{P}{\rho} = RT$$

So,

$$\frac{\Delta P}{\Delta \rho} : \frac{P}{\rho} = RT \quad \text{or} \quad \left(\frac{\Delta P}{\Delta \rho}\right)_T = \frac{P}{\rho} = RT$$

INTRODUCTION TO FLUID KINEMATICS

Thus, **bulk modulus** of ideal gas can be express as

$$k = \rho\left(\frac{\Delta P}{\Delta \rho}\right) = \rho\left(\frac{P}{\rho}\right) = P$$

$$\Rightarrow k = P$$

Therefore, coefficient of compressibility of an ideal gas is equivalent to its absolute pressure. The expression indicates that compressibility of the gas will be increase on increasing the absolute pressure. Also, for an ideal gas, we have

$$\frac{\Delta P}{\Delta \rho} : \frac{P}{\rho} = RT \Rightarrow \frac{\Delta P}{P} = \frac{\Delta \rho}{\rho}$$

thus, fractional changes in the density and the pressure of an ideal gas are equivalent in magnitude. The increment in pressure of the gas during isothermal compression results into the increasing its density. The coefficient of compressibility is given by

$$k \cong -\frac{\Delta P}{\Delta V/V} \approx \frac{\Delta P}{\Delta \rho/\rho}$$

Where $\Delta V/V$ or $\Delta \rho/\rho$ is dimensionless, then k is of same dimension of pressure i.e., Pascal (Pa). Also, mathematical expression of bulk modulus shows that compressibility of a fluid depends on the variation of the pressure with relation to the fractional change of volume (or density) at the constant temperature (T). For example, the pressure of water at normal atmospheric condition increased up-to 210 atm causes a change of 1% in its density corresponding to a coefficient of compressibility value of $k = 21,000$ atm. Thus, high variation in pressure brings only differential changes in volume (or density) of a liquid with relation to higher value of **Bulk Modulus** (k).

Let us consider the case of gases involve in air at 1 atm pressure, $k = P = 1$ atm and a decrease of 1% in its density $(\Delta \rho/\rho = 0.01)$ results in increases the pressure $(\Delta P = 0.01 \text{ atm})$. Again, air at 10000 atm and a decrease of 1% in density results in increases the pressure up-to

INTRODUCTION TO FLUID KINEMATICS

$\Delta P = 100$ atm. Thus, marginal variation in density of a gas results in a large variation in pressure at constant temperature.

The inverse of the co-efficient of compressibility is known as **isothermal compressibility** α which is defined as

$$\alpha = \frac{1}{k} = -\frac{1}{V}\left(\frac{\Delta V}{\Delta P}\right)_T = \frac{1}{\rho}\left(\frac{\Delta \rho}{\Delta P}\right)_T$$

COMPRESSION BY TEMPERATURE

The volume (or density) strongly depends on the temperature of the surrounding. It has observed that the volume (or density) of many liquids increases on raising the temperature. To quantify these effects, we need an attribute which represents the changes in volume (or density) in relation to the variation of temperature at constant pressure. The **co-efficient of volume expansion** signify the fractional changes in volume (or density) due to the variation in temperature at constant pressure. Mathematically, it can expresses as

$$\beta = \frac{1}{V}\left(\frac{\Delta V}{\Delta T}\right)_P = -\frac{1}{\rho}\left(\frac{\Delta \rho}{\Delta T}\right)_P$$

$$\Rightarrow \beta = \frac{\Delta V/V}{\Delta T} = -\frac{\Delta \rho/\rho}{\Delta T}$$

For Ideal gas, and from the equation of state, we obtain as
$P = \rho RT \Rightarrow \Delta P = \Delta \rho R \Delta T$

$$\Rightarrow \frac{\Delta P}{P} = \frac{\Delta \rho R \Delta T}{\rho RT} = \frac{\Delta \rho \Delta T}{\rho T}$$

Since pressure is constant $\therefore \Delta P = P$, then we get

$$\frac{\Delta P}{P} = \frac{P}{P} = 1 = \frac{\Delta \rho \Delta T}{\rho T}$$

$$\Rightarrow \frac{\Delta \rho}{\rho} = \frac{T}{\Delta T}$$

$$\therefore \beta = T$$

INTRODUCTION TO FLUID KINEMATICS

Where T is absolute temperature.

The large value of co-efficient of volume expansion β indicates large variation in density due to high temperature. For example, winds blow due to natural convection where density difference is directly proportional to fractional changes in temperature of surrounding at constant pressure. So, strong convective currents pass through air at higher temperature.

The combined effects of pressure and temperature on the volume (or density) of a fluid can be calculated from the specific volume in terms of T and P. Consider the specific volume as the function of T and P which can be written as
$V = V(T,P)$

Differentiating V and express in terms of α and β, we get

$$dV = \left(\frac{\partial V}{\partial T}\right)_P dT + \left(\frac{\partial V}{\partial P}\right)_T dP$$
$$\Rightarrow dV = (\alpha dT - \beta dP)V$$
$$\Rightarrow \frac{dV}{V} = \alpha dT - \beta dP$$

STEADY & UNSTEADY FLOW

The term steady indicates that properties of a fluid remain constant at any fixed point with time. The reverse happening considered as unsteady. During the steady flow, the fluid properties such as velocity, temperature can vary from one point to another but they remain unchanged at any particular point. Thus, the volume, density, total energy of a steady flow through turbines, pumps, boilers, heat exchangers of power plants or compressors remain constant with time. Some devices such as reciprocating engines do not suitable for steady flow as the flow through inlet and outlet are not steady. Moreover, the fluid properties change over a time period, then these flow can be study based on time-averaged methods. Computational techniques are being used to analyze the unsteady fluid flow which involves flow induced vibrations, pressure fluctuations, turbulent eddies, and shock waves.

INTRODUCTION TO FLUID KINEMATICS

Mathematically, we can expresses the steady flow as

$$\left(\frac{\Delta \rho}{\Delta t}\right)_{x_1,y_1,z_1} = k_1, \left(\frac{\Delta V}{\Delta t}\right)_{x_1,y_1,z_1} = k_2, \left(\frac{\Delta P}{\Delta t}\right)_{x_1,y_1,z_1} = k_3$$

Thus, steady flow is defined as the fluid flow whose properties such as density, volume, pressure, etc. remain constant with time at any particular point. In same way, unsteady flow is defined as the fluid flow whose properties such as density, volume, pressure, etc. vary with time at any particular point. Mathematically, it can be expresses as

$$\left(\frac{\Delta \rho}{\Delta t}\right)_{x_1,y_1,z_1} \neq 0, \left(\frac{\Delta V}{\Delta t}\right)_{x_1,y_1,z_1} \neq 0, \left(\frac{\Delta P}{\Delta t}\right)_{x_1,y_1,z_1} \neq 0$$

INTERNAL & EXTERNAL FLOW

A fluid flow is classified as internal or external flow based on the flow past through bounded or open surfaces. **Internal flows** can be considered as those flow through bounded area such as flow through pipes and ducts, etc. These flows are greatly influenced by the viscosity or Reynolds number. On the other side, **External flows** are those fluid flow past through open surface e.g., flow over the plate and flow of water in lakes. Moreover, open-channel flow is the subcategory of the external flow which includes flow through opened duct or irrigation ditches.

Internal flows governs by pressure difference provided other attributes are least significant. Moreover, dimension of inlet portion of pipes and ducts are effect on the fluid flow. This pressure difference has correlation with other characteristics of the flows which differentiate it into laminar or turbulent flows. Usually, flow through circular pipes are considered as internal flow as it is convenient to assume fully developed laminar flow through circular pipe for theoretical study.

External flow is defined as the flow over bodies which are immersed in a fluid. Here, the pressure force exerted normal to the surface of the bodies and shear forces acts parallel to the surface of the bodies. The component of the resultant force which acts on the flow direction known as drag forces and the component which acts normal to the flow direction is known as lift forces. External flows are governs by drag forces and lift forces. The skin friction and coefficient of drag and lift of a flow are important attributes while studying external flows. The viscous effects are less effective to the flow region such as shock waves, flow over airfoils, flow over cylinders and spheres.

LAMINAR & TURBULENT FLOW

It has been observed from the flame of burner that flame rises smoothly for the first few centimeters and then starts fluctuating randomly as long it rise up as shown in figure.

Here, portion of flame which rise up smoothly is known as laminar, characterized by smooth streamlines and ordered movement of fluid. The upper most portion of flame moves randomly considered to be turbulent characterized by random streamlines and disordered movement.

The fluid elements moves in straight line or fluid layers are parallel to each other so that fluid elements do not cross each other lines. Laminar flow is possible only at low fluid motion and with high viscosity. The fluid elements cannot move in straight lines in less viscous fluid which makes turbulent flow with high fluid motion. The turbulent flow in which the fluid elements moves in zig-zag way. Most of the flows encountered in our daily life are turbulent. However, highly viscous fluid such as engine oil, dyes flow through narrow pipe considered as laminar flow.

INTRODUCTION TO FLUID KINEMATICS

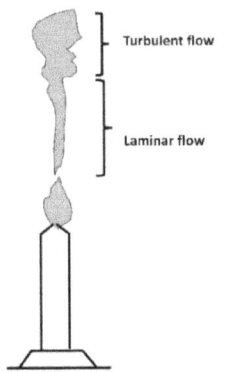

We can verify the existence of the laminar, transitional, and turbulent flow regimes by injecting dye streaks into the flow in a glass pipe, as the British engineer Osborne Reynolds did over double century ago. We observe that dye streak shows smooth line at low fluid motion indicates laminar flow, followed by few fluctuations indicates transitional regime, and zig-zag randomly lines indicates turbulent flow. This high fluctuation in flow is mainly due to the rapid mixing of fluid elements from the adjacent layers of dye.

ROTATIONAL & IRROTATIONAL FLOW

The fluid flows classified into **rotational and irrotational flow** based on its angular movement. When the fluid elements of any fluid flow moves in streamlines, also rotates about their own axis, such flow known as **rotational flow.** While, the fluid elements of any fluid flow moves in streamlines but do not rotate about their own axis, such flow is known as **irrotational flow.**

If **q** be the velocity vector of a fluid particle, then the vector quantity, Ω ($= curl\ q$), is called the *vorticity vector* or simply the *vorticity* and is a measure of the angular velocity of an infinitesimal element.

Let $\Omega = \Omega_x i + \Omega_y j + \Omega_z k$ so that ($\Omega_x, \Omega_y, \Omega_z$) are the *vorticity components* or the *components of the spin*. Then, if $q = u\,i + v\,j + wk$.

$$\Omega_x = \frac{\partial w}{\partial y} - \frac{\partial v}{\partial z} \qquad \Omega_y = \frac{\partial u}{\partial z} - \frac{\partial w}{\partial x} \qquad \Omega_z = \frac{\partial v}{\partial x} - \frac{\partial u}{\partial y}$$

If $\Omega_x, \Omega_y, \Omega_z$, are all zero, the motion is *irrotational* and the velocity function ϕ exists and if $\Omega_x, \Omega_y, \Omega_z$ are not all zero, the motion is *rotational*. Thus, the fluid flow which possess irrotational motion considered as **irrotational flows,** while the flow which possess rotational motion considered as **rotational flows.**

ONE, TWO, & THREE DIMENSIONAL FLOW

INTRODUCTION TO FLUID KINEMATICS

A fluid flow is classified based on the dimensional distribution of its velocity. It can be one, two, three dimensional velocity distribution respectively. In general velocity profile are being study under two dimensional geometry. However, in advance study, a three dimensional geometry being used for the velocity distribution of a fluid flow such as $v(x \ y \ z)$ in rectangular dimensions, $v(r \ \theta \ z)$ in cylindrical coordinates, and $v(r \ \theta \ \phi)$ in polar coordinates. The relative variation of velocity w.r.t. other directions are negligible then it is prefer to choose one, two dimensional study. The velocity distribution is uniform across the pipe, since the velocity is nearly constant at all radii except very close to the wall of the body. For example, the flow of air over a car antenna preferred as two-dimensional except near its end as the airflow over car antenna is uniform. Also, the velocity distribution of a fluid flows through circular pipe is two-dimensional as the velocity shows the variation in $r-$ direction and $z-$ direction but not in $\theta-$ direction.

UNIFORM & NON-UNIFORM FLOW

The flow in which the fluid motion does not vary with space (i.e., length of direction of the flow) at fixed time is called as **uniform flow.** On other side, the flow in which the fluid motion does vary with space at fixed time is called as **non-uniform flow.** Mathematically, uniform flow and non-uniform flow can be express as

$$\left(\frac{\Delta v}{\Delta l}\right)_t = \text{constant} \quad \text{and} \quad \left(\frac{\Delta v}{\Delta l}\right)_t \neq \text{constant}$$

NEWTONIAN & NON-NEWTONIAN FLOW

A fluid which obeys the Newton's law of viscosity is referred as **Newtonian flow.** The Newtonian flow is such a fluid flow where shear stress is proportional to the velocity gradient. For example, water obey Newton's law of viscosity under the normal conditions. These fluid flows in relative motion where internal resistance induced

VISCOUS & NON-VISCOUS FLOW

An internal resistance induced between the layers of a fluid

INTRODUCTION TO FLUID KINEMATICS

flow when their layers moves relatively. It is known as viscosity which measures of internal resistance ability of a fluid. Viscosity is basically due to cohesive forces between the molecules of the fluid. Hypothetically, we consider an ideal fluid is the fluid having no viscosity while all fluids have viscosity to some extent. The fluid flow which possess significant level of viscosity referred to be as **viscous flow.** However, there are regions (close to surface) where viscous forces are insignificant, referred as **inviscid flow regions.** The region of fluid flow on surface with thin boundary layer are unaffected by inertial or pressure forces is known as **inviscid flow region.**

NATURAL & FORCED FLOW

The fluid flow can be classified based on the applied force. When a fluid flows under the action of any applied forces by external means such as pumps or compressor, then such flow called as **forced flow.** On the other side, if any fluid flow under the action of buoyancy forces, such flow called as **natural flow.** For example, warmer winds blows to less denser regions.

METHODS OF DESCRIBING FLUID MOTION

There are two methods for studying fluid motion mathematically. These are Lagrangian and Eulerian methods and refer to 'individual time-rate of change' and 'local time rate of change' respectively.

LAGRANGIAN METHOD

In this method we study the history of each fluid particle, i.e., any fluid particle is selected and is pursued on its onward course observing the changes in the velocity, pressure and density at each point and at each instant. Let (x_o, y_o, z_o) be the coordinates of the chosen particle at a given time $t = t_o$. At a later time, $t = t$, let the coordinates of the same particle be (x, y, z). Since the chosen particle is any particle in the fluid, the coordinates (x, y, z) will be functions of t and also of their

initial values (x_0, y_0, z_0), so that
$$x = f_1(x_o, y_o, z_o, t)$$

INTRODUCTION TO FLUID KINEMATICS

$$y = f_2(x_o, y_o, z_o, t)$$
$$z = f_3(x_o, y_o, z_o, t) \quad ----(1)$$

Let u, v, w and a_x, a_y, a_z be the components of velocity and acceleration respectively. Then, we have

$$u = \frac{\partial x}{\partial t}, \quad v = \frac{\partial y}{\partial t}, \quad w = \frac{\partial z}{\partial t} \quad -------(2)$$

And

$$a_x = \frac{\partial^2 x}{\partial t^2}, \quad a_y = \frac{\partial^2 y}{\partial t^2}, \quad a_z = \frac{\partial^2 z}{\partial t^2} \quad ------(3)$$

So, in the Lagrangian description of fluid flow, individual fluid particles are considered and their positions, velocities, etc. are described as a function of time. As the particle move in the flow field, their positions and velocities change with time. The physical laws, such as Newton's laws and conservation of mass and energy, apply directly to each particle. If there were only a few particles to consider, as in a high school physics experiment, the Lagrangian description would be desirable. However, fluid flow is a continuum phenomenon, at least down to the molecular level. It is not possible to track each particle in a complex flow field. Thus, the Lagrangian description is rarely used in fluid mechanics.

Remark 1. The fundamental equation of motion in Lagrangian form are non-linear and hence it leads to many difficulties while solving a problem. In fact, the present method is employed with an advantage only in some one - dimensional problems. Hence, we need to think about another method of describing fluid motion.

Remark 2. This method resembles that of dynamics of a particle in so far as (x, y, z) are dependent on t. Thus, in Lagrangian method of fluid dynamics (x, y, z) are dependent on four independent variables x_o, y_o, z_o, t.

EULERIAN METHOD

In this method we select any point fixed in space occupied by the fluid and study the changes which take place in velocity, pressure and density as the fluid passes through this point.

INTRODUCTION TO FLUID KINEMATICS

$$u = F_1(x,y,z,t) \quad v = F_2(x,y,z,t) \quad w = F_3(x,y,z,t) \quad \text{-----(4)}$$

For a particular value of t, (4) exhibits the motion at all points in the fluid at that time. Again, for a particular point (x,y,z), u, v, w are functions of t, which define the mode of variations of velocity at that point.

Remark 1. The point under consideration being fixed, x,y,z and t are independent variables and hence $\dfrac{dx}{dt}, \dfrac{d^2x}{dt^2}$ etc. have no meaning in this method.

However, the Eulerian description is one in which a control volume is defined, within which fluid flow properties of interest are expressed as fields. Pressure, velocity, acceleration, and all other flow properties are described as fields within the control volume. In other words, each property is expressed as a function of space and time. Here, velocity, acceleration, etc. of whatever particle happens to be at a particular location of interest at a particular time. Since fluid flow is a continuum phenomenon, atleast down to the molecular level, the Eulerian description is usually preferred in fluid mechanics.

Some translation or reformulation of these laws is required for use with a Eulerian description.

1. **Pressure field**: An example of a fluid flow variable expressed in Eulerian terms is the pressure. Rather than the pressure of an individual particle, a pressure field is introduced, i.e., $p = p(x,y,z,t)$

Note that pressure is a scalar, and is written as a function of space and time (x, y,z and t). In other words, at a given point in space (x, y, and z), and at some particular time (t), the pressure is defined. In the Eulerian description, it is of no concern which fluid particle is at that location at that time. In fact, whatever fluid particle happens to be at that location at time t experiences the pressure defined above.

2. **Velocity field**: An example of a fluid flow variable expressed in Eulerian terms is the velocity. Rather than following the velocity of an individual particle, a velocity field is introduced, i.e.,

$$\vec{V}(x,y,z,t) = u(x,y,z,t)\hat{i} + v(x,y,z,t)\hat{j} + w(x,y,z,t)\hat{k}$$

INTRODUCTION TO FLUID KINEMATICS

Note that since velocity is a vector, it can be split into three components (u, v, and w), all three of which are functions of space and time (x, y, z, and t). In other words, at a given point in space (x, y, and z), and at some particular time (t), the velocity vector is defined. In the Eulerian description, it is of no concern which fluid particle is at that location at that time. In fact, whatever fluid particle happens to be at that location at that time. In fact, whatever fluid particle happens to be at that location at time t has the velocity defined above.

3. **Acceleration field**: An example of a fluid flow variable expressed in Eulerian terms is the acceleration. Instead of this, acceleration of an individual particle, an acceleration field is introduced as

$$\vec{a}(x,y,z,t) = \vec{a_x}(x,y,z,t)\hat{i} + \vec{a_y}(x,y,z,t)\hat{j} + \vec{a_z}(x,y,z,t)\hat{k}$$

Note that since acceleration is a vector, it can be split into three components, all three of which are functions of space and time (x, y, z, and t). In other words, at a given point in space (x, y, z, and t), and at some particular time (t), the acceleration vector is defined. In the Eulerian description, it is of no concern which fluid particle is at that location at that time. In fact, whatever fluid particle happens to be at that location at time t has the acceleration defined above. Thus, the Eulerian description is usually preferred because there are simply too many particles to keep track of in a Lagrangian description.

MATERIAL, LOCAL AND CONVECTIVE DERIVATIVE

According to the Lagrangian method, the equation of motion for any fixed fluid particle can be formulated, where fluid particle moves in a flow define as material particle. Thus, we can define a fluid particle's position in space in terms of a material position vector as $(x_{particle}(t), y_{particle}(t), z_{particle}(t))$. We know that the acceleration of the fluid particle is the time derivative of the fluid particle's velocity which can be express as

INTRODUCTION TO FLUID KINEMATICS

$$\vec{a} = \frac{d\vec{v}_{particle}}{dt}$$

The acceleration of the particle is equivalent to the standardized value of given acceleration field at the position $(x_{particle}(t), y_{particle}(t), z_{particle}(t))$ of particular fluid particle at any instant time t. In other words, at any instant time t, the acceleration field $\vec{a} = \vec{a}(x, y, z, t)$ is equivalent to the acceleration of the fluid particle positioned at (x,y,z). Thus, for a control volume system, we can transform fundamental equations from the Lagrangian to the Eulerian frame of reference for improved analysis of fluid's attributes. The motion of a fluid's particle at the material position vector is defined as

$$\vec{V}_{particle}(t) = \vec{V}\left(x_{particle}(t), y_{particle}(t), z_{particle}(t), t\right)$$

Since the dependent variable $\vec{V}$ is a function of four independent variables $x_{particle}, y_{particle}, z_{particle}$ and t. We apply chain rule to determine the time derivative of velocity field as

$$\vec{a}_{particle} = \frac{d\vec{V}}{dt} = \frac{d\vec{V}_{particle}}{dt} = \frac{d\vec{V}\left(x_{particle}, y_{particle}, z_{particle}, t\right)}{dt}$$

$$\Rightarrow \vec{a}_{particle} = \frac{\delta\vec{V}}{\delta t}\frac{dt}{dt} + \frac{\delta\vec{V}}{\delta x_{particle}}\frac{dx_{particle}}{dt} + \frac{\delta\vec{V}}{\delta y_{particle}}\frac{dy_{particle}}{dt} + \frac{\delta\vec{V}}{\delta z_{particle}}\frac{dz_{particle}}{dt}$$

With Lagrangian frame of reference, at any instant time t, we have

$$\frac{dx_{particle}}{dt} = u \text{ and } \frac{dy_{particle}}{dt} = v \text{ and } \frac{dz_{particle}}{dt} = w \text{ and }$$
$$\frac{dt}{dt} = 1$$

The acceleration field w.r.t. Eulerian frame of reference can be express as

$$\vec{a}(x, y, z, t) = \frac{d\vec{V}}{dt} = \frac{\partial\vec{V}}{\partial t} + u\frac{\partial\vec{V}}{\partial x} + v\frac{\partial\vec{V}}{\partial y} + w\frac{\partial\vec{V}}{\partial z}$$

INTRODUCTION TO FLUID KINEMATICS

$$\Rightarrow \vec{a}(x,y,z,t) = \frac{d\vec{V}}{dt} = \frac{\partial \vec{V}}{\partial t} + (\vec{V}.\vec{\nabla})\vec{V}$$

Where $\vec{\nabla}$ is the del operator, vector operator, which is define in coordinate system as

$$\vec{\nabla} = \frac{\partial}{\partial x}i + \frac{\partial}{\partial y}j + \frac{\partial}{\partial z}k$$

Thus, components of acceleration of a fluid's particle is defined as

$$a_x = \frac{\partial u}{\partial t} + u\frac{\partial u}{\partial x} + v\frac{\partial u}{\partial y} + w\frac{\partial u}{\partial z}$$

$$a_y = \frac{\partial v}{\partial t} + u\frac{\partial v}{\partial x} + v\frac{\partial v}{\partial y} + w\frac{\partial v}{\partial z}$$

$$a_z = \frac{\partial w}{\partial t} + u\frac{\partial w}{\partial x} + v\frac{\partial w}{\partial y} + w\frac{\partial w}{\partial z}$$

In the above equations, the expression $\frac{\partial u}{\partial t}, \frac{\partial v}{\partial t}, \frac{\partial w}{\partial t}$ are called **local acceleration**. It is defined as the rate of increment in fluid motion w.r.t. time at a particular point in space or control volume. The expression other than $\frac{\partial u}{\partial t}, \frac{\partial v}{\partial t}, \frac{\partial w}{\partial t}$ in the above equations are called as **convective acceleration**. It is defined as the rate of change of fluid's motion due to the change of position of fluid's particles in a fluid flow field. The terms a_x, a_y and a_z are the **total acceleration** in the x,y and z directions respectively. Thus, total derivative operator called as **material derivative** symbolize as D/Dt which is defined as

$$\frac{D}{Dt} \equiv \frac{d}{dt} = \frac{\partial}{\partial t} + (\vec{V}.\vec{\nabla})$$

It is clearly understood that the **material derivative** is equivalent to the sum of **local** and **convective** components of respective field.

INTRODUCTION TO FLUID KINEMATICS

Generally, material derivative of an acceleration field can be expressed as

$$\Rightarrow \vec{a}(x,y,z,t) = \frac{D\vec{V}}{Dt} \equiv \frac{d\vec{V}}{dt} = \frac{\partial \vec{V}}{\partial t} + (\vec{V}.\vec{\nabla})\vec{V}$$

Note that **material derivative** can be apply to different fluid's attributes including both scalar and vectors. Thus, material derivative of a pressure field (scalar quantity) can expressed as

$$P(x,y,z,t) \equiv \frac{D(P)}{Dt} = \frac{\partial P}{\partial t} + (\vec{V}.\vec{\nabla})P$$

Let us derive the expression for the material derivative of a scalar and vector quantity characterized in a fluid flow field.

Suppose a fluid particle moves from $P(x,y,z)$ at time t to $Q(x + \delta x, y + \delta y, z + \delta z)$ at time $t + \delta t$. Furthermore, suppose $f(x, y, z, t)$ be **scalar function** associated with some property of the fluid. Let the total change of f due to movement of the fluid particle from P to Q be δf. Then, we have

$$\delta f = \left(\frac{\delta f}{\delta x}\right)\delta x + \left(\frac{\delta f}{\delta y}\right)\delta y + \left(\frac{\delta f}{\delta z}\right)\delta z + \left(\frac{\delta f}{\delta t}\right)\delta t$$

$$\Rightarrow \frac{\delta f}{\delta t} = \frac{\partial f}{\partial x}\frac{\delta x}{\delta t} + \frac{\partial f}{\partial y}\frac{\delta y}{\delta t} + \frac{\partial f}{\partial z}\frac{\delta z}{\delta t} + \frac{\partial f}{\partial t} \quad ----(1)$$

Let

$$\lim_{\delta t \to 0} \frac{\delta f}{\delta t} = \frac{Df}{Dt} \text{ or } \frac{df}{dt}$$

$$\lim_{\delta t \to 0} \frac{\delta x}{\delta t} = \frac{dx}{dt} = u$$

$$\lim_{\delta t \to 0} \frac{\delta y}{\delta t} = \frac{dy}{dt} = v$$

$$\lim_{\delta t \to 0} \frac{\delta z}{\delta t} = \frac{dz}{dt} = w$$

Where $q = (u,v,w)$ is the velocity of the fluid particle at P.

INTRODUCTION TO FLUID KINEMATICS

Making $\delta t \to 0$ and using above relations, (1) reduces to

$$\frac{Df}{Dt} = u\frac{\partial f}{\partial x} + v\frac{\partial f}{\partial y} + w\frac{\partial f}{\partial z} + \frac{\partial f}{\partial t} \quad \text{---------- (2)}$$

But $q = u\,\mathbf{i} + v\,\mathbf{j} + w\,\mathbf{k}$ \quad ---------(3)

and $\nabla = \frac{\partial}{\partial x}\mathbf{i} + \frac{\partial}{\partial y}\mathbf{j} + \frac{\partial}{\partial z}\mathbf{k}$ \quad ---------(4)

From (3) and (4), $q.\nabla = u\frac{\partial}{\partial x} + v\frac{\partial}{\partial y} + w\frac{\partial}{\partial z}$ ----(5)

Using (5), the equation (2) reduces to

$$\frac{Df}{Dt} = \frac{\partial f}{\partial t} + (q.\nabla)f \quad \text{-----(A)}$$

Similar again suppose a fluid particle moves from $P(x, y, z)$ at time t to $Q(x + \delta x, y + \delta y, z + \delta z)$ at time $t + \delta t$. Furthermore, suppose $g(x, y, z, t)$ be **vector function** associated with some property of the fluid. Let the total change of g due to movement of the fluid particle from P to Q be δg. Then, we have

$$\delta g = \left(\frac{\delta g}{\delta x}\right)\delta x + \left(\frac{\delta g}{\delta y}\right)\delta y + \left(\frac{\delta g}{\delta z}\right)\delta z + \left(\frac{\delta g}{\delta t}\right)\delta t$$

$$\Rightarrow \frac{\delta g}{\delta t} = \frac{\partial g}{\partial x}\frac{\delta x}{\delta t} + \frac{\partial g}{\partial y}\frac{\delta y}{\delta t} + \frac{\partial g}{\partial z}\frac{\delta z}{\delta t} + \frac{\partial g}{\partial t} \quad \text{----- (6)}$$

Let

$$\lim_{\delta t \to 0} \frac{\delta g}{\delta t} = \frac{Dg}{Dt} \text{ or } \frac{dg}{dt}$$

$$\lim_{\delta t \to 0} \frac{\delta x}{\delta t} = \frac{dx}{dt} = u$$

$$\lim_{\delta t \to 0} \frac{\delta y}{\delta t} = \frac{dy}{dt} = v$$

$$\lim_{\delta t \to 0} \frac{\delta z}{\delta t} = \frac{dz}{dt} = w$$

Where $q = (u, v, w)$ is the velocity of the fluid particle at P. Making $\delta t \to 0$ and using above

relations, (6) reduces to
$$\frac{Dg}{Dt} = u\frac{\partial g}{\partial x} + v\frac{\partial g}{\partial y} + w\frac{\partial g}{\partial z} + \frac{\partial g}{\partial t} \quad ----(7)$$

But $q = u\,\boldsymbol{i} + v\,\boldsymbol{j} + w\,\boldsymbol{k}$ -------------(8)

and $\nabla = \frac{\partial}{\partial x}\boldsymbol{i} + \frac{\partial}{\partial y}\boldsymbol{j} + \frac{\partial}{\partial z}\boldsymbol{k}$ -----------(9)

From (8) and (9), $q.\nabla = u\frac{\partial}{\partial x} + v\frac{\partial}{\partial y} + w\frac{\partial}{\partial z}$ ----(10)

Using (10), the equation (7) reduces to

$$\frac{Dg}{Dt} = \frac{\partial g}{\partial t} + (q.\nabla)g \quad -----(B)$$

From (A) and (B), we have for both scalar and vector functions

$$\frac{D}{Dt} = \frac{\partial}{\partial t} + q.\nabla \quad --(11)$$

$\frac{D}{Dt}$ is called the **material** (*or particle or substantial*) **derivative**. The first term on R.H.S. of (11), namely $\frac{\partial}{\partial t}$, is called the **local derivative** and it is associated with time variation at a fixed position. The second term on R.H.S. of (11), namely $q.\nabla$, is called the **convective derivative** and it is associated with the change of a physical quantity f or g due to motion of the fluid particle.

Example 1: Acceleration of a fluid particle

Suppose a fluid particle moves from $P(x, y, z)$ at time t to $Q(x + \delta x, y + \delta y, z + \delta z)$ at time $t + \delta t$. Furthermore, suppose $v(x, y, z, t)$ be velocity (vector) function associated with some property of the fluid. Let the total change of f due to movement of the fluid particle from P to Q be δf. Then, we have

$$\delta\vec{v} = \left(\frac{\delta\vec{v}}{\delta x}\right)\delta x + \left(\frac{\delta\vec{v}}{\delta y}\right)\delta y + \left(\frac{\delta\vec{v}}{\delta z}\right)\delta z + \left(\frac{\delta\vec{v}}{\delta t}\right)\delta t$$

$$\Rightarrow \frac{\delta\vec{v}}{\delta t} = \frac{\partial\vec{v}}{\partial x}\frac{\delta x}{\delta t} + \frac{\partial\vec{v}}{\partial y}\frac{\delta y}{\delta t} + \frac{\partial\vec{v}}{\partial z}\frac{\delta z}{\delta t} + \frac{\partial\vec{v}}{\partial t} \quad -----(1) \qquad \text{Let}$$

INTRODUCTION TO FLUID KINEMATICS

$$\lim_{\delta t \to 0} \frac{\delta \vec{v}}{\delta t} = \frac{D\vec{v}}{Dt} \text{ or } \frac{d\vec{v}}{dt}$$

$$\lim_{\delta t \to 0} \frac{\delta x}{\delta t} = \frac{dx}{dt} = u$$

$$\lim_{\delta t \to 0} \frac{\delta y}{\delta t} = \frac{dy}{dt} = v$$

$$\lim_{\delta t \to 0} \frac{\delta z}{\delta t} = \frac{dz}{dt} = w$$

Where $q = (u,v,w)$ is the velocity of the fluid particle at P. Making $\delta t \to 0$ and using above relations, (1) reduces to

$$\frac{D\vec{v}}{Dt} = u\frac{\partial \vec{v}}{\partial x} + v\frac{\partial \vec{v}}{\partial y} + w\frac{\partial \vec{v}}{\partial z} + \frac{\partial \vec{v}}{\partial t} \quad \text{------(2)}$$

But $q = u\,\vec{i} + v\,\vec{j} + w\,\vec{k}$ ------------(3)

and $\nabla = \frac{\partial}{\partial x}\vec{i} + \frac{\partial}{\partial y}\vec{j} + \frac{\partial}{\partial z}\vec{k}$ ------------(4)

From (3) and (4), $q.\nabla = u\frac{\partial}{\partial x} + v\frac{\partial}{\partial y} + w\frac{\partial}{\partial z}$ ----(5)

Using (5), the equation (2) reduces to

$$\vec{a} = \frac{D\vec{v}}{Dt} \text{ or } \frac{d\vec{v}}{dt} = \frac{\partial \vec{v}}{\partial t} + (q.\nabla)\vec{v} \quad \text{-----(C)}$$

Example 2: Pressure of a fluid particle

Suppose a fluid particle moves from $A(x,y,z)$ at time t to $B(x+\delta x, y+\delta y, z+\delta z)$ at time $t+\delta t$. Furthermore, suppose $P(x,y,z,t)$ be pressure (scalar) function associated with some property of the fluid. Let the total change of P due to movement of the fluid particle from A to B be δf. Then, we have

$$\delta P = \left(\frac{\delta P}{\delta x}\right)\delta x + \left(\frac{\delta P}{\delta y}\right)\delta y + \left(\frac{\delta P}{\delta z}\right)\delta z + \left(\frac{\delta P}{\delta t}\right)\delta t$$

$$\Rightarrow \frac{\delta P}{\delta t} = \frac{\partial P}{\partial x}\frac{\delta x}{\delta t} + \frac{\partial P}{\partial y}\frac{\delta y}{\delta t} + \frac{\partial P}{\partial z}\frac{\delta z}{\delta t} + \frac{\partial P}{\partial t} \quad \text{------ (1)}$$

Let

INTRODUCTION TO FLUID KINEMATICS

$$\lim_{\delta t \to 0} \frac{\delta P}{\delta t} = \frac{DP}{Dt} \text{ or } \frac{dP}{dt}$$

$$\lim_{\delta t \to 0} \frac{\delta x}{\delta t} = \frac{dx}{dt} = u$$

$$\lim_{\delta t \to 0} \frac{\delta y}{\delta t} = \frac{dy}{dt} = v$$

$$\lim_{\delta t \to 0} \frac{\delta z}{\delta t} = \frac{dz}{dt} = w$$

Where $q = (u,v,w)$ is the velocity of the fluid particle at A. Making $\delta t \to 0$ and using above relations, (1) reduces to

$$\frac{DP}{Dt} = u\frac{\partial P}{\partial x} + v\frac{\partial P}{\partial y} + w\frac{\partial P}{\partial z} + \frac{\partial P}{\partial t} \quad --------- (2)$$

But $q = u\,\mathbf{i} + v\,\mathbf{j} + w\,\mathbf{k}$ ------------(3)

and $\nabla = \frac{\partial}{\partial x}\mathbf{i} + \frac{\partial}{\partial y}\mathbf{j} + \frac{\partial}{\partial z}\mathbf{k}$ -----------(4)

From (3) and (4), we have

$$q.\nabla = u\frac{\partial}{\partial x} + v\frac{\partial}{\partial y} + w\frac{\partial}{\partial z} \quad -----(5)$$

Using (5), the equation (2) reduces to

$$\frac{DP}{Dt} = \frac{\partial P}{\partial t} + (q.\nabla)P -----(D)$$

FLOW VISUALIZATION

Streamlines

Streamlines are widely used indicator for the fluid flow visualization in a particular fluid field. A streamline define as a tangent to the velocity vector curve of a fluid element in a particular flow field. Streamlines cannot be directly observed experimentally except in few cases.

INTRODUCTION TO FLUID KINEMATICS

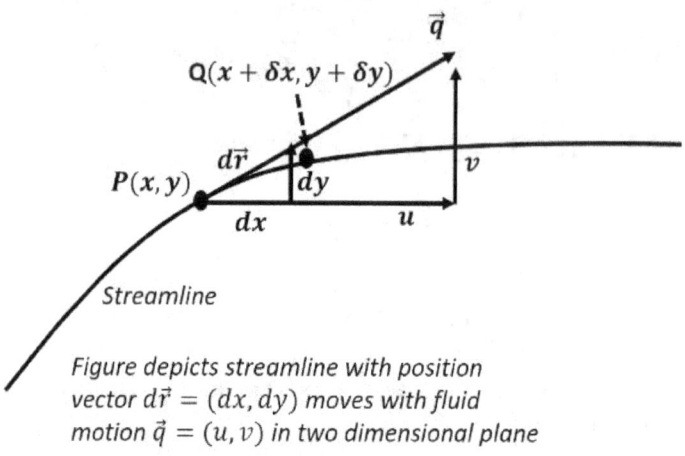

Figure depicts streamline with position vector $d\vec{r} = (dx, dy)$ moves with fluid motion $\vec{q} = (u, v)$ in two dimensional plane

Mathematically, we can express streamlines of the flow field as follow:

Let us assume infinitesimal arc length given by

$$d\vec{r} = dx\,\hat{i} + dy\,\hat{j} + dz\,\hat{k}$$

Where $d\vec{r}$ is parallel to the velocity vector which is given by
$\vec{V} = u\hat{i} + v\hat{j} + w\hat{k}$

From the figure, it is cleared that the components of $d\vec{r}$ are proportional to the components of $\vec{V}$ using similar triangles properties and we get

$$\frac{dr}{V} = \frac{dx}{u} = \frac{dy}{v} = \frac{dz}{w}$$

Where dr is the magnitude of $\vec{dr}$ and V is the magnitude of velocity vector $\vec{V}$. We can obtain equation of streamlines in two dimensions using the equation as mention below:

$$\frac{dr}{V} = \frac{dx}{u} = \frac{dy}{v}$$

On solving the above equations numerically, we obtain a set of family curves with an arbitrary constant of integration. Thus, each chosen value of constant of integration represents a particular streamline and the family of curves represents streamlines of a fixed flow field.

Pathlines

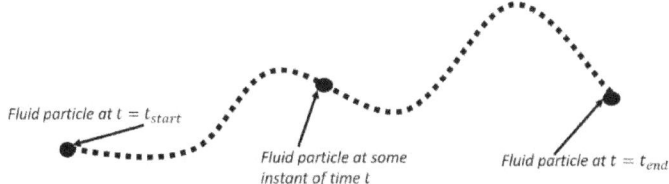

According to Lagrangian theory, a pathline is the locus of a fluid particle over stipulated time period. Pathlines are widely used to analysis the fluid flow. Thus, a pathline is defined as the fluid's particle movement with material position vector $\left[x_{particle}(t), y_{particle}(t), z_{particle}(t)\right]$ over some time period. The figure shows the curve followed by an individual fluid particle for a certain time period, $t_{start} < t < t_{end}$, where fluid particle seems to be popping up and down like an ocean waves at the beach.

Recently, a new experimental method known as **Particle image velocimetry (PIV)** used to study the fluid particle's pathlines over an entire flow field.

Mathematically, it can be expresses as

$$\frac{dr}{dt} = V$$

$$\Rightarrow \frac{dx}{dt} = u, \quad \frac{dy}{dt} = v, \quad \frac{dz}{dt} = w$$

Where
$$V = u\hat{i} + v\hat{j} + w\hat{k}$$

And
$$r = x\hat{i} + y\hat{j} + z\hat{k}$$

Note: Let a fluid particle of fixed identity be at (x_o, y_o, z_o) when $t = t_o$, then the path lines are determined from the equations

$$\frac{dx}{dt} = u(x, y, z, t)$$

INTRODUCTION TO FLUID KINEMATICS

$$\frac{dy}{dt} = v(x,y,z,t)$$

$$\frac{dz}{dt} = w(x,y,z,t)$$

With initial conditions

$$x(t_o) = x_o, \quad y(t_o) = y_o, \quad z(t_o) = z_o$$

Streaklines

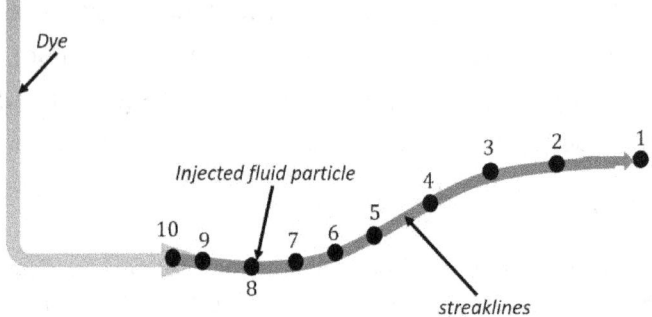

A streak line is a line on which all those fluid particle lies that are at some instant passed through a certain point of space. Thus, a streak line represents the instantaneous pictures of the position of all fluid particles, which have passed through a given point at some previous time. When a dye is injected into a moving fluid at some fixed point, the visible lines produced in the fluid are streak lines which have passed through the injected point as shown in figure.

Note: The streamlines, pathlines, and streaklines are identical when the fluid flow is steady.

The equation of the streak line at time t can be derived by Lagrangian method, suppose that a fluid particle $(x_o, y_o, z_o,)$ passes a fixed point (x_1, y_1, z_1) in the course of time. Then by using the Lagrangian method of description, we have

$$f_1(x_o,y_o,z_o,t) = x_1, \quad f_2(x_o, y_o, z_o,t) = y_1,$$
$$f_3(x_o, y_o, z_o, t) = z_1$$

INTRODUCTION TO FLUID KINEMATICS

solving above equation for x_o, y_o, z_o, we have

$x_o = g_1(x_1, y_1, z_1, t)$, $y_o = g_2(x_1, y_1, z_1, t)$, $z_o = g_3(x_1, y_1, z_1, t)$

Now a streak line is the locus of the positions (x,y,z) of the particles which have passed through the fixed point (x_1, y_1, z_1).

Hence the equation of the streak line at time t is given by

$x = h_1(x_o, y_o, z_o, t)$, $\quad y = h_2(x_o, y_o, z_o, t)$, $\quad z = h_3(x_o, y_o, z_o, t)$

Substituting the values of x_o, y_o, z_o in above equation, the required equation of streak line passing through (x_1, y_1, z_1) at time t is given by

$x = h_1(g_1, g_2, g_3, t)$, $\quad y = h_2(g_1, g_2, g_3, t)$, $\quad z = h_3(g_1, g_2, g_3, t)$

Timelines

A timeline is set of tracing lines of fluid particles which are observed at instant of time. In order to check the uniformity of a fluid flow, timelines are being used. The timelines are used to create velocity vector plot at some particular time. Timelines can be generated experimentally in a water by using a hydrogen bubble wire. To conduct electrolysis of the water, a small voltage of current is sent through the cathode wire as a result tiny hydrogen gas bubbles are produced in the wire. Due to small size of hydrogen bubbles, these bubbles are following the flow pattern and can used to create the timelines for different instant of time.

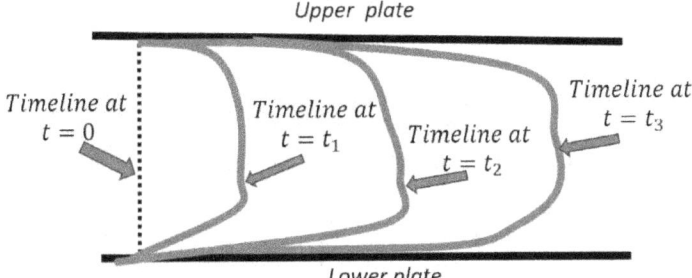

Figure depicts the timelines in a channel flow past through two parallel plates. Due to the friction at surface of the plates, the fluid motion is almost zero (considered as no-slip

INTRODUCTION TO FLUID KINEMATICS

condition), and top and bottom of the timeline are marked from their starting positions. In the regions of flow away from the plate's surface, the marked fluid particles moves with the average fluid speed which deforms the timelines.

VELOCITY POTENTIAL FUNCTION

Suppose that the fluid velocity at time t is $q = (u,v,w)$

Further suppose that at the considered instant t, there exists a scalar function $\phi(x, y, z, t)$, uniform throughout the entire field of flow and such that

$$-d\phi = udx + vdy + wdz$$

$$\Rightarrow \left(\frac{\partial \phi}{\partial x} dx + \frac{\partial \phi}{\partial y} dy + \frac{\partial \phi}{\partial z} dz\right) = udx + vdy + wdz$$

$$u = -\frac{\partial \phi}{\partial x}, \quad v = -\frac{\partial \phi}{\partial y}, \quad w = -\frac{\partial \phi}{\partial z}$$

$$\Rightarrow q = -\nabla.\phi = -grad\ \phi$$

ϕ is called *velocity potential*. The negative sign indicates that the flow takes place from the higher to lower potentials.

EQUIPOTENTIALS

The surfaces $\phi(x, y, z, t) = constants$ are *equipotential*.

The streamlines are given by

$$\frac{dx}{u} = \frac{dy}{v} = \frac{dz}{w}$$

Are cut at right angles by the surfaces given by the DE

$$udx + vdy + wdz = 0$$

Proof: Suppose that the fluid velocity at time t is $q = (u, v, w)$

Further suppose that at the considered instant t, there exists a scalar function $\phi(x, y, z, t)$, uniform throughout the entire field of flow and such that

$$-d\phi = udx + vdy + wdz$$

INTRODUCTION TO FLUID KINEMATICS

$$\Rightarrow \left(\frac{\partial \phi}{\partial x}dx + \frac{\partial \phi}{\partial y}dy + \frac{\partial \phi}{\partial z}dz\right) = udx + vdy + wdz$$

$$u = -\frac{\partial \phi}{\partial x}, \quad v = -\frac{\partial \phi}{\partial y}, \quad w = -\frac{\partial \phi}{\partial z}$$

The analytical condition being

$$u\left(\frac{\partial w}{\partial y} - \frac{\partial v}{\partial z}\right) + v\left(\frac{\partial u}{\partial z} - \frac{\partial w}{\partial x}\right) + w\left(\frac{\partial v}{\partial x} - \frac{\partial u}{\partial y}\right) = 0 - -(A)$$

When the velocity potential exists, then

$$\frac{\partial w}{\partial y} - \frac{\partial v}{\partial z} = -\frac{\partial^2 \phi}{\partial y \partial z} + \frac{\partial^2 \phi}{\partial z \partial y} = 0 \Rightarrow \frac{\partial w}{\partial y} = \frac{\partial v}{\partial z}$$

$$\frac{\partial u}{\partial z} - \frac{\partial w}{\partial x} = -\frac{\partial^2 \phi}{\partial z \partial x} + \frac{\partial^2 \phi}{\partial x \partial z} = 0 \Rightarrow \frac{\partial u}{\partial z} = \frac{\partial w}{\partial x}$$

$$\frac{\partial v}{\partial x} - \frac{\partial u}{\partial y} = -\frac{\partial^2 \phi}{\partial x \partial y} + \frac{\partial^2 \phi}{\partial y \partial x} = 0 \Rightarrow \frac{\partial v}{\partial x} = \frac{\partial u}{\partial y}$$

Using the above relations, condition given by equ (4) is satisfied.

Hence surface exist which cut the streamlines orthogonally. So, we concluded that all points of field of flow, the equipotential are cut orthogonally by the streamlines.

STREAM FUNCTION

Consider the continuity equation of an incompressible, two-dimensional flow in cartesian coordinates system as

$$\frac{\partial u}{\partial x} + \frac{\partial v}{\partial y} = 0$$

On introducing stream function (ψ), we can transformed the continuity equation in terms of one dependent variable (ψ) instead of two dependent variables $(u \text{ and } v)$. Thus, we have

INTRODUCTION TO FLUID KINEMATICS

$$\frac{\partial}{\partial x}\left(\frac{\partial \psi}{\partial y}\right) + \frac{\partial}{\partial y}\left(-\frac{\partial \psi}{\partial x}\right) = \frac{\partial^2 \psi}{\partial x \partial y} - \frac{\partial^2 \psi}{\partial y \partial x} = 0$$

Where stream function defined as

$$u = \frac{\partial \psi}{\partial y} \quad \text{and} \quad v = -\frac{\partial \psi}{\partial x}$$

Also, stream function for a compressible, two-dimensional flow in cartesian coordinates is given by

$$\rho u = \frac{\partial \psi}{\partial y} \quad \text{and} \quad \rho v = -\frac{\partial \psi}{\partial x}$$

Also, stream function for an incompressible, two-dimensional flow in cylindrical coordinates is given by

$$u_r = -\frac{1}{r}\frac{\partial \psi}{\partial z} \quad \text{and} \quad u_z = \frac{1}{r}\frac{\partial \psi}{\partial r} \quad \text{and in polar coordinates,}$$

$$u_r = \frac{1}{r}\frac{\partial \psi}{\partial \theta} \quad \text{and} \quad u_\theta = -\frac{\partial \psi}{\partial r}$$

Note: signs convention of stream function are relevant. The $-ve$ sign for v chosen arbitrarily as the fluid flows from left to right such that ψ increases in the $y-axis$. However, some fluid mechanics books, signs convention for ψ are opposite. (e.g., in some British reference books and book written by Heinsohn and Cimbala, 2003).

BOOSTER CAPSULE : MACH NUMBER

The gases flow through compressors (or orifices) are considered as compressible flow where density of the fluid vary throughout the flow. Moreover, the density vary from point to point in compressible flow. Thus, variation of density of a fluid depends on the variation in pressure and temperature.

It is a common observation that a fluid (liquid or gas) expands or contracts with a variation of pressure acts on it. Fluids compress when pressurized and expands in volume (or density) when heated or depressurized. The proportion of compression of fluids is vary for different fluids. So, we can say that fluids behave like elastic solids w.r.t. pressure. This compression of fluids is mainly due to the pressure waves which produce a disturbance in a fluid flow. However, this disturbance propagating with speed of sound in flow field. The disturbance creates in a solids, liquid or gas is propagating from one point to another. The speed with which the disturbance is propagating depends upon the molecular distance of state of matter. As molecules in solids are closely packed and thus the disturbance is propagating instantaneously. But this is not in the case of fluids (liquids and gases) where molecules are relatively apart. Thus, each molecule covered a certain distance before it will propagating the disturbance, and as result the speed of propagation of disturbance in fluids are comparatively less than in case of solids. Hence, the speed of propagation of disturbance depends on the pressure and density of the fluids as molecular distance is corelated with density.

When any kind of disturbance is produced in an incompressible fluid, then it is propounding in all directions with the speed of sound (c) and it is created due to the pressure difference induced in a fluid which flow with the speed (v). The behaviour of the propagation of disturbance depends on the **Mach number**. Let us discuss in details:

CASE 1: $M<1$ We know that Mach number is defined as

$$\text{Mach number} = \sqrt{\frac{\text{Inertia force}}{\text{Elastic force}}} = \sqrt{\frac{\rho a v^2}{\kappa a}} = \sqrt{\frac{\rho v^2}{\kappa}} = \sqrt{\frac{v^2}{\kappa/\rho}} = \frac{v}{\sqrt{\kappa/\rho}} = \frac{v}{c}$$

$$\Rightarrow \text{Mach number} = \frac{v}{c} = \frac{\text{fluid motion}}{\text{speed of sound}}$$

In this case, we have Mach number $(M) = \frac{v}{c} < 1 \Rightarrow v < c.$

Which means that the speed of the moving fluid (v) is less than speed of sound (c) and this type of flow is known as sub-sonic flow. In order to observe the behaviour of the propagation of disturbance for this case, represent the propagation graphically as follow:

INTRODUCTION TO FLUID KINEMATICS

Consider $v=1$ unit and $c=3$ units, $\dfrac{v}{c}=\dfrac{1}{3}<1$. Assume that propagation is at point P and is moving towards right such that it reaches the point Q in 4 seconds.

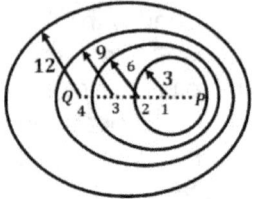

Figure visualize propagation of pressure wave when $M < 1$

The position of the propagation after 1 sec, 2 sec, 3 sec and 4 sec along the lines are shown by the mark 4,3, 2 and 1 respectively. The propagation moves from P to B in 4 seconds and hence the distance $PQ = 4 \times v = 4 \times 1 = 4$ units. The disturbance produced at P in 4 seconds will move a distance $= 4 \times c = 4 \times 3 = 12$ units in all directions. Thus, taking P as centre and radius equal to 12 units, a circle is drawn. This circle gives the position of disturbance after 4 seconds. When the propagation is at mark 3, it will reach Q in 3 seconds and distance and distance covered $= 3 \times v = 3 \times 1 = 3$ units. Moreover, disturbance will move a distance having radius $= 3 \times c = 3 \times 3 = 9$ units. In same way, at mark 2, the disturbance will have a radius $= 2 \times c = 2 \times 3 = 6$ units and at mark 1, the disturbance will have a radius $= 1 \times c = 1 \times 3 = 3$ units. As shown in figure, the disturbance produced due to the pressure wave is ahead of the propagation of fluid and point Q is inside the sphere of radius 12 units in the case when $v < c$.

CASE 2: $M = 1$ In this case, Mach number $(M) = \dfrac{v}{c} = 1 \Rightarrow v = c$ which means that speed of the flowing fluid (v) is equivalent to the speed of sound (c) in magnitude as well as in direction and this type of flow is known as sonic flow. Let us understand the behaviour of the propagation of disturbance for this case when plotted graphically as follow:

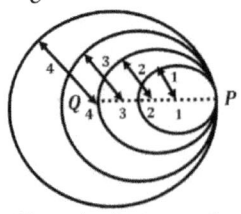

Figure visualize propagation of pressure wave when $M = 1$

INTRODUCTION TO FLUID KINEMATICS

Assume that the fluid flows from point P to Q in 4 seconds. The disturbance produced at point P will move a distance having radius $= 4 \times c = 4 \times 1 = 4$ units in every directions. Thus, taking P as centre and radius equal to 4 units, a circle is drawn. This circle gives the position of disturbance after 4 seconds. In the same way, disturbance produced at the mark 3 will covered distance having radius $= 3 \times c = 3 \times 1 = 3$ units in every directions and draw a circle whose centre as mark 3 and radius equal to 3 units. This circle gives the position of disturbance after 3 seconds. Similarly, disturbance produced at the mark 2 and 1 covered distance having radius 2 and 1 respectively. This is shown in figure.

CASE 3: $M > 1$ In this case, Mach number $(M) = \dfrac{v}{c} > 1 \Rightarrow v > c$ which means that speed of the flowing fluid is much greater than the speed of sound travelled by propagation and this type of flow is known as supersonic flow. In order to observe the behaviour of the propagation of disturbance for this case, represent the propagation graphically as follow:

Take $v = 1$ unit and $c = 0.4$, then $M = \dfrac{v}{c} = \dfrac{1}{0.4} = 2.5 > 1$.

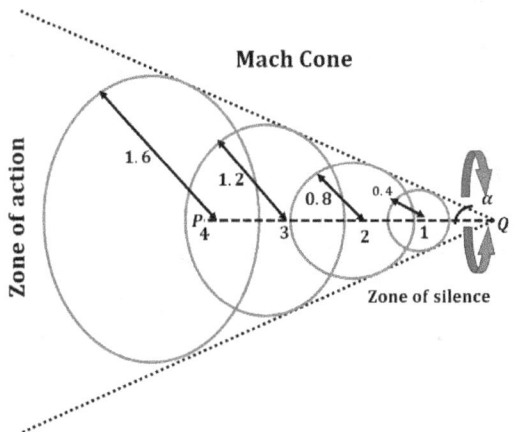

Assume that the propagation moves from P to Q in 4 seconds. The distance travelled by the flowing fluid in 4 seconds $= 4 \times v = 4 \times 1 = 4$ units. Thus, consider $PQ = 4$

units. The disturbance produced at P will move in all directions having radius $= 4 \times c = 4 \times 0.4 = 1.6$ units. Take P as centre, draw a circle with the radius equal to 1.6 units. After one second, the propagation will be at mark 3 and distance covered $= 3 \times v = 3 \times 1 = 3$ units and disturbance produced at mark 3 will move in all directions having radius $= 3 \times c = 3 \times 0.4 = 1.2$ units. In the same way, disturbance produced at mark 2 and 1 will covered distance having radius $= 2 \times c = 2 \times 0.4 = 0.8$ unit and $1 \times c = 1 \times 0.4 = 0.4$ unit as shown in figure. It can be observed that sphere of propagation of disturbance is lag behind the projection of flowing fluid. If we draw a tangent to the different circles on both sides, we get a cone with vertex at Q. This cone is called as Mach Cone.

Let α be half of the angle of the Mach cone, which called as Mach angle and is given by

$$\sin \alpha = \frac{1R}{1B} = \frac{c}{v} = \frac{1}{v/c} = \frac{1}{M}$$

Also, disturbance has been felt only in the region covered Mach cone when $M > 1$. This region is considered as action zone. The region exterior to Mach cone considered as silence zone where no disturbance has been felt.

Q. Prove that the stream function (ψ) is constant for a set of streamlines of the flow?

Solution: The equation of a streamline for an incompressible ,two-dimensional flow is given by

$$\frac{dx}{u} = \frac{dy}{v} \Rightarrow -vdx + udy = 0 \quad \text{-------(1)}$$

On introducing stream function , the equation (1) reduced to

$$\frac{\partial \psi}{\partial x}dx + \frac{\partial \psi}{\partial y}dy = 0 \quad \text{------(2)}$$

Moreover, the stream function ψ can be expresses in terms of two independent variables x and y as :

INTRODUCTION TO FLUID KINEMATICS

$$d\psi = \frac{\partial \psi}{\partial x}dx + \frac{\partial \psi}{\partial y}dy \quad \text{------(3)}$$

From equation (2) and (3), we have $d\psi = 0$ along a streamline, thus it shows that the stream function ψ is constant along streamlines of any fluid flow.

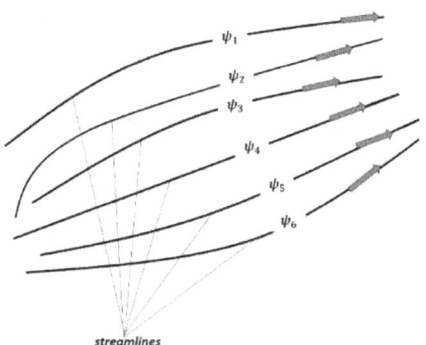

streamlines

Consider two-dimensional fluid flow in xy-plane, the collection of curves of constant obtain from the differential equation $d\psi = 0$ represents the streamlines of the flow. Thus, for each chosen value of ψ_i, the stream function represent streamline in radial direction and a family of curves represents streamlines as shown in figure.

HIGHER ORDER THINKING SKILL QUESTION

❖ **ILLUSTRATION OF MATERIAL DERIVATIVE**

CASE STUDY: A fluid flow field is given by

$$V = (yz+t)\hat{i} + (xz-t)\hat{j} + xy\hat{k}$$

Determine the acceleration at the point $(1,2,3)$ at $t = 1.5$ sec Also, Sketch the material acceleration vectors at the various array of x- and y-values.

Solution: Velocity V of the given fluid flow at point (x, y, z) is given by

INTRODUCTION TO FLUID KINEMATICS

$$V = u\hat{i} + v\hat{j} + w\hat{k} = (yz+t)\hat{i} + (xz-t)\hat{j} + xy\hat{k}$$

$\Rightarrow u = yz + t \qquad \therefore \dfrac{\partial u}{\partial x} = 0$

$\qquad v = xz - t \qquad \therefore \dfrac{\partial v}{\partial y} = 0$

$\qquad w = xy \qquad \therefore \dfrac{\partial w}{\partial z} = 0$

For a case of possible fluid flow, the continuity equation must be satisfied.

$$\dfrac{\partial u}{\partial x} + \dfrac{\partial v}{\partial y} + \dfrac{\partial w}{\partial z} = 0$$

Put the values of $\dfrac{\partial u}{\partial x}, \dfrac{\partial v}{\partial y}$ and $\dfrac{\partial w}{\partial z}$, we get

$$\dfrac{\partial u}{\partial x} + \dfrac{\partial v}{\partial y} + \dfrac{\partial w}{\partial z} = 0 + 0 + 0 = 0$$

The velocity field $V = (yz+t)\hat{i} + (xz-t)\hat{j} + xy\hat{k}$ is a possible fluid flow.

The acceleration components a_x, a_y and a_z for the given flow field are

$$a_x = \dfrac{Du}{Dt} = \dfrac{\partial u}{\partial t} + u\dfrac{\partial u}{\partial x} + v\dfrac{\partial u}{\partial y} + w\dfrac{\partial u}{\partial z}$$

$$a_y = \dfrac{Dv}{Dt} = \dfrac{\partial v}{\partial t} + u\dfrac{\partial v}{\partial x} + v\dfrac{\partial v}{\partial y} + w\dfrac{\partial v}{\partial z}$$

$$a_z = \dfrac{Dw}{Dt} = \dfrac{\partial w}{\partial t} + u\dfrac{\partial w}{\partial x} + v\dfrac{\partial w}{\partial y} + w\dfrac{\partial w}{\partial z}$$

The components $u, v,$ and w of velocity are given by

INTRODUCTION TO FLUID KINEMATICS

$u = yz + t \quad \therefore \dfrac{\partial u}{\partial t} = 1, \dfrac{\partial u}{\partial x} = 0, \dfrac{\partial u}{\partial y} = z, \dfrac{\partial u}{\partial z} = y$

$v = xz - t \quad \therefore \dfrac{\partial v}{\partial t} = -1, \dfrac{\partial v}{\partial x} = z, \dfrac{\partial v}{\partial y} = 0, \dfrac{\partial v}{\partial z} = x$

$w = xy \quad \therefore \dfrac{\partial w}{\partial t} = 0, \dfrac{\partial w}{\partial x} = y, \dfrac{\partial w}{\partial y} = x, \dfrac{\partial w}{\partial z} = 0$

Put these values in acceleration components, we get acceleration at $(1,2,3)$ at $t = 1.5$ sec

$a_x = \dfrac{\partial u}{\partial t} + u\dfrac{\partial u}{\partial x} + v\dfrac{\partial u}{\partial y} + w\dfrac{\partial u}{\partial z}$

$\Rightarrow a_x = 1 + (yz+t) \times 0 + (xz-t) \times z + (xy) \times y$

$\Rightarrow a_x = 1 + 0 + (1 \times 3 - 1.5) \times 3 + (1 \times 2) \times 2$

$\Rightarrow a_x = 1 + 0 + 4.5 + 4 = 9.5$ units

$a_y = \dfrac{\partial v}{\partial t} + u\dfrac{\partial v}{\partial x} + v\dfrac{\partial v}{\partial y} + w\dfrac{\partial v}{\partial z}$

$\Rightarrow a_y = -1 + (yz+t) \times z + (xz-t) \times 0 + (xy) \times x$

$\Rightarrow a_y = -1 + (2 \times 3 + 1.5) \times 3 + 0 + (1 \times 2) \times 1$

$\Rightarrow a_y = -1 + 22.5 + 0 + 2 = 23.5$ units

$a_z = \dfrac{\partial w}{\partial t} + u\dfrac{\partial w}{\partial x} + v\dfrac{\partial w}{\partial y} + w\dfrac{\partial w}{\partial z}$

$\Rightarrow a_z = 0 + (yz+t) \times y + (xz-t) \times x + (xy) \times 0$

$\Rightarrow a_z = 0 + (2 \times 3 + 1.5) \times 2 + (1 \times 3 - 1.5) \times 1 + 0$

$\Rightarrow a_z = 0 + 15 + 1.5 + 0 = 16.5$ units

$\therefore$ Acceleration $= a_x \hat{i} + a_y \hat{j} + a_z \hat{k} = 9.5\hat{i} + 23.5\hat{j} + 16.5\hat{k}$

The acceleration of the given fluid flow is given by

INTRODUCTION TO FLUID KINEMATICS

$$a = \sqrt{a_x^2 + a_y^2 + a_z^2} = \sqrt{(9.5)^2 + (23.5)^2 + (16.5)^2}$$
$$\Rightarrow a = \sqrt{90.25 + 552.25 + 272.25} = \sqrt{914.75}$$
$$\Rightarrow a = 30.25 \text{ units}$$

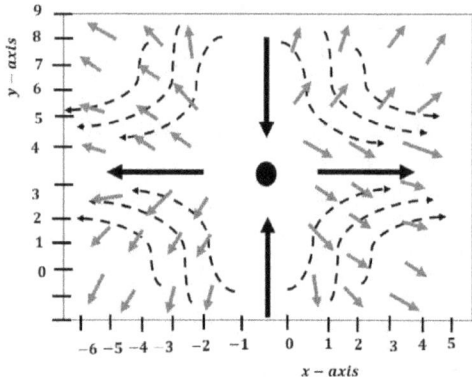

The acceleration field is nonzero as the flow is unsteady. Above the stagnation point $(y = 1.87)$, the acceleration vectors plotted in figure pointed upward, increasing in magnitude away from the stagnation point. To the right of the stagnation point, the acceleration vectors point to the right, again increasing in magnitude away from the stagnation point. This agrees qualitatively with the velocity vectors and streamlines. The fluid particles in the upper-right of the flow field are accelerated in upper-right direction and moves in counterclockwise direction due to centripetal acceleration toward the upper right. The flow below $y = 1.87$ is the mirror image of the flow above the symmetry line, and the flow to the left of $x = -0.76$ is a mirror of the flow to the right of the symmetry line.

CASE STUDY: The fluid flows through a pipe at $0.01 \text{ m}^3/\text{sec}$ and pipe converges uniformly from 0.5 m to 0.3 m in diameter over 3 m long pipe. If the rate of flow changes from $0.01 \text{ m}^3/\text{sec}$ to $0.03 \text{ m}^3/\text{sec}$, in 30 sec, calculate the material acceleration at the middle of the pipe after 16 seconds.

Solution: According to Law of mass conservation, the rate of discharge is constant and equal to $0.01 \text{ m}^3/\text{sec}$. In this case, the fluid motion is in x-direction only. Thus, this is a one-dimensional flow and velocity components in y and z direction are zero (i.e., $v = 0, w = 0$).

$$\therefore \quad \text{Convective acceleration} = u\frac{\partial u}{\partial x} \text{ only}$$

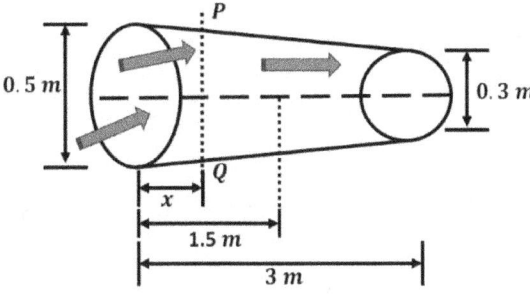

The diameter (D_i) at a distance x from the inlet is given by

$$D_i = D_1 - \frac{D_1 - D_2}{l} \times x$$

$$\Rightarrow D_i = 0.5 - \frac{0.5 - 0.3}{3} \times x$$

$$\Rightarrow D_i = (0.5 - 0.07x) \text{ m}$$

The area of cross-section (A_{PQ}) at section PQ is given by

$$A_{PQ} = \frac{\pi}{4}D_i^2 = \frac{\pi}{4}(0.5 - 0.07x)^2$$

Velocity (u) at the section PQ in terms of Q (i.e., in terms of rate of flow)

INTRODUCTION TO FLUID KINEMATICS

$$u = \frac{Q}{\text{Area}} = \frac{Q}{A_{PQ}} = \frac{Q}{\frac{\pi}{4}D_i^2}$$

$$\Rightarrow u = \frac{4Q}{\pi(0.5-0.07x)^2}$$

$$\Rightarrow u = 1.273Q(0.5-0.07x)^{-2} \text{ m/sec} \quad \text{------(1)}$$

The rate of flow at $t = 16$ seconds is given by

$$Q = Q_1 + \frac{Q_2 - Q_1}{30} \times 16$$

$$\Rightarrow Q = 0.01 + \frac{0.03 - 0.01}{30} \times 16$$

$$\Rightarrow Q = (0.01 + 0.01067) \text{ m}^3/\text{sec}$$

$$\Rightarrow Q = 0.02067 \text{ m}^3/\text{sec}$$

To find $\frac{\partial u}{\partial t}$, we must differentiate equation (1) with respect to t.

$$\therefore \quad \text{Local acceleration} = \frac{\partial u}{\partial t} = \frac{\partial}{\partial t}\left[1.273Q(0.5-0.07x)^{-2}\right]$$

$$= 1.273 \times (0.5-0.07x)^{-2} \times \frac{\partial Q}{\partial t}$$

[∵ Local acceleration is at a point where x is constant but Q is changing]

Local acceleration at the middle of the pipe (at $x = 1.5$ m)

INTRODUCTION TO FLUID KINEMATICS

$$= 1.273 \times (0.5 - 0.07 \times 1.5)^{-2} \times \frac{\partial Q}{\partial t}$$

$$= 1.273 \times \frac{1}{0.156025} \times \frac{\partial Q}{\partial t}$$

$$= 8.16 \times \frac{0.02}{30} \quad \left[\because \frac{\partial Q}{\partial t} = \frac{Q_2 - Q_1}{t} = \frac{0.03 - 0.01}{30} = \frac{0.02}{30} \right]$$

$$= 0.00544 \text{ m/s}^2$$

To find $\frac{\partial u}{\partial x}$, we must differentiate equation (1) with respect to x.

$$\frac{\partial u}{\partial x} = \frac{\partial}{\partial x} \left[1.273 Q (0.5 - 0.07x)^{-2} \right]$$

$$\Rightarrow \frac{\partial u}{\partial x} = 1.273 Q (-2)(0.5 - 0.07x)^{-3}(-0.07)$$

$$\Rightarrow \frac{\partial u}{\partial x} = 0.17822 Q (0.5 - 0.07x)^{-3}$$

$\therefore$ Convective acceleration $= u \frac{\partial u}{\partial x}$

$$= \left[1.273 Q (0.5 - 0.07x)^{-2} \right] \left[0.17822 Q (0.5 - 0.07x)^{-3} \right]$$

$$= 1.273 \times 0.17822 \times Q^2 \times (0.5 - 0.07x)^{-5}$$

$$= 1.273 \times 0.17822 \times (0.02067)^2 \times (0.5 - 0.07x)^{-5}$$

Convective acceleration at the middle of the pipe $(\text{at } x = 1.5 \text{ m})$

$$= 1.273 \times 0.17822 \times (0.02067)^2 \times (0.5 - 0.07 \times 1.5)^{-5}$$

$$= 0.34 \text{ m/s}^2$$

$\therefore$ Material acceleration = Convective acceleration + Local acceleration

$$= 0.34 + 0.00544 = 0.34544 \text{ m/s}^2$$

❖ Illustration of compressibility

INTRODUCTION TO FLUID KINEMATICS

CASE STUDY OF COMPRESSION THROUGH TEMPERATURE

Consider water at $35°C$ and 1 atm. Determine the change in its density when it is heated up-to $70°C$ provided pressure remains constant. It is given that volume expansivity, $\beta = 0.337 \times 10^{-3} K^{-1}$.

Analysis: To determine the variation of density of the water on heating, we need to assume following assumptions as:

1. The coefficient of volume expansion and the isothermal compressibility of water are constant in the given temperature.
2. The approximate analysis is performed by replacing differential changes in quantities by fractional changes.
3. Initial density of the water before heated taken as 998 kg/m³.

We knew that the effects of pressure and temperature on the density, ρ of any fluid can be determined by taking specific density which is function of temperature, T and pressure P. thus, variation of volume of any fluid is given by

$$d\rho = \left(\frac{\partial \rho}{\partial T}\right)_P dT + \left(\frac{\partial \rho}{\partial P}\right)_T dP$$

$$\Rightarrow d\rho = (-\beta dT + \alpha dP)\rho \quad \left[\because \alpha = \frac{1}{\rho}\left(\frac{\partial \rho}{\partial T}\right)_P \text{ and } \beta = -\frac{1}{\rho}\left(\frac{\partial \rho}{\partial P}\right)_T\right]$$

$$\Rightarrow \frac{d\rho}{\rho} \approx \frac{\Delta \rho}{\rho} = \alpha dP - \beta dT$$

$$\Rightarrow \frac{\Delta \rho}{\rho} \equiv \alpha \Delta P - \beta \Delta T$$

$$\Rightarrow \Delta \rho = \alpha \rho \Delta P - \beta \rho \Delta T \quad \text{------(1)}$$

Where α = thermal compressibility of water and
β = coefficient of volume expansion of water at absolute

INTRODUCTION TO FLUID KINEMATICS

temperature

In this case, the pressure remains constant throughout the heating process. So, we have $\Delta P = 0$ and equation (1) reduce to

$$\Delta \rho = -\beta \rho \Delta T \quad \text{------(2)}$$

As per question,

$\beta = 0.337 \times 10^{-3} \; K^{-1}$

$\rho = 998 \; kg/m^3$

$\Delta T = T_{final} - T_{initial} = 70°C - 35°C = 35°C = 35 + 273 = 308 \; K$

Put these values in equation (2), we get

$\Delta \rho = -(0.337 \times 10^{-3}).(998).(308)$

$\Rightarrow \Delta \rho = \rho_{final} - \rho_{initial} = -103.6 \; kg/m^3$

$\Rightarrow \rho_{final} = 103 + \rho_{initial} = -103.6 + 998 = 894.4 \; kg/m^3$

CASE STUDY OF COMPRESSION THROUGH PRESSURE

Consider water at $35°C$ and 1 atm. Determine the change in its density when it is compressed to 101 atm pressure provided temperature remains constant. Use the value of isothermal compressibility of water, $\alpha = 4.80 \times 10^{-5} \; atm^{-1}$.

To determine the variation of density of the water on compression, we need to assume following assumptions as:

1. The coefficient of volume expansion and the isothermal compressibility of water are constant in the given temperature.
2. The approximate analysis is performed by replacing differential changes in quantities by fractional changes.
3. Initial density of the water before heated taken as $998 \; kg/m^3$.

We knew that the effects of pressure and temperature on the

INTRODUCTION TO FLUID KINEMATICS

density, ρ of any fluid can be determined by taking specific density which is function of temperature, T and pressure P. thus, variation of volume of any fluid is given by

$$d\rho = \left(\frac{\partial \rho}{\partial T}\right)_P dT + \left(\frac{\partial \rho}{\partial P}\right)_T dP$$

$$\Rightarrow d\rho = (-\beta dT + \alpha dP)\rho \quad \left[\because \alpha = \frac{1}{\rho}\left(\frac{\partial \rho}{\partial T}\right)_P \text{ and } \beta = -\frac{1}{\rho}\left(\frac{\partial \rho}{\partial P}\right)_T\right]$$

$$\Rightarrow \frac{d\rho}{\rho} \approx \frac{\Delta \rho}{\rho} = \alpha dP - \beta dT$$

$$\Rightarrow \frac{\Delta \rho}{\rho} \equiv \alpha \Delta P - \beta \Delta T$$

$$\Rightarrow \Delta \rho = \alpha \rho \Delta P - \beta \rho \Delta T \quad \text{------(1)}$$

Where α = thermal compressibility of water and

β = coefficient of volume expansion of water at absolute temperature

In this case, the temperature remains constant throughout the compression process. So, we have $\Delta T = 0$ and equation (1) reduce to

$$\Delta \rho = \alpha \rho \Delta P \quad \text{------(2)}$$

As per question,

$\alpha = 4.80 \times 10^{-5}$ atm^{-1}

$\rho = 998$ kg/m^3

$\Delta P = P_{final} - P_{initial} = 101 - 1 = 100$ atm

Put these values in equation (2), we get

$$\Delta \rho = (4.80 \times 10^{-5}).(998).(100)$$

$$\Rightarrow \Delta \rho = \rho_{final} - \rho_{initial} = 4.8 \text{ kg/m}^3$$

$$\Rightarrow \rho_{final} = 4.8 + \rho_{initial} = 4.8 + 998 = 1002.8 \text{ kg/m}^3$$

<u>Conclusion:</u> It is clearly observed that density of water

INTRODUCTION TO FLUID KINEMATICS

decreases on heating and increases when compressed.

❖ PRACTICAL APPROACH OF STREAM FUNCTION

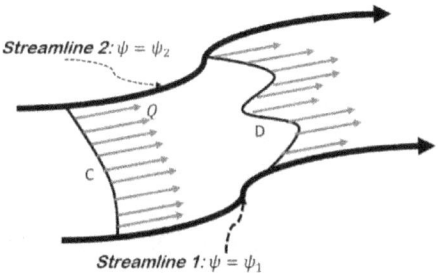

The figure depicts two streamlines ψ_1 and ψ_2 of two-dimensional flow in the XY-plane, of unit width into the page (1 m in z-direction). On applying law of conservation of mass to the given flow, we can say that no flow can cross any streamline or in other words, fluid which occupy the space between two streamlines remains confined between them. It means that the volume flow rate through any cross-sectional area between streamlines remains constant at any instant of time. Here, cross-sectional area may be of any dimensions, provided that it starts at streamline 1 and ends at streamline 2. In figure, cross-sectional area C is smooth arc of the streamline 1 while cross-sectional D is wavy of streamline 2 where discharge rate Q (volume flow rate) of steady, incompressible, two-dimensional flow remains constant. Moreover, the distance between two cross-sectional area increases then fluid motion between the two streamlines decreases accordingly. Similarly, when streamline converges then fluid motion between the two streamlines will increases. Let us prove the fact that *discharge rate from streamline 1 to streamline 2 is equal to the difference between the values of the two stream functions.*

INTRODUCTION TO FLUID KINEMATICS

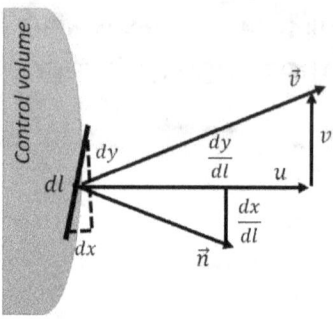

Consider a control volume bounded by the two streamlines as shown in figure. Let dl be the length of cross-sectional portion D of streamline 2 along its unit normal vector $\vec{n}$.

Moreover, unit normal vector is defined by

$$\vec{n} = \frac{dy}{dl}\hat{i} - \frac{dx}{dl}\hat{j}$$

Discharge rate through portion dl of the control volume is

$$dQ = \vec{v}.\vec{n}.\underbrace{dA}_{dl} = \left(u\hat{i} + v\hat{j}\right)\left(\frac{dy}{dl}\hat{i} - \frac{dx}{dl}\hat{j}\right)dl \quad \text{-----(1)}$$

Applying vector algebra, equation (1) reduced to

$$\Rightarrow dQ = u\,dy - v\,dx$$

$$\Rightarrow dQ = \frac{\partial \psi}{\partial y}dy + \frac{\partial \psi}{\partial x}dx \quad \left[\because u = \frac{\partial \psi}{\partial y} \text{ and } v = -\frac{\partial \psi}{\partial x}\right]$$

$$\Rightarrow dQ = d\psi$$

In order to calculate the total discharge rate through cross-sectional portion D, we must integrate above expression from streamline 1 to streamline 2,

$$Q = \int_D \vec{v}.\vec{n}.dA = \int_D dQ = \int_{\psi=\psi_1}^{\psi=\psi_2} d\psi = \psi_2 - \psi_1$$

Since we choose cross-sectional portion D is of any shape or position within the space confined between two streamlines, thus, the statement is proven.

CASE STUDY: A fluid is created a sink flow through a narrow

INTRODUCTION TO FLUID KINEMATICS

orifice on the bottom of a container. The fluid flows from left to right with the velocity $v = 3 \text{ m}^2/\text{s}$.

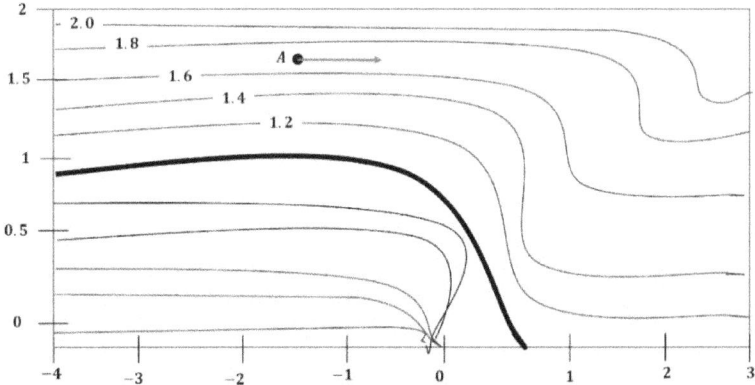

The water channel is perpendicular to the xy-plane, and runs along the z-axis across the entire flow which has width $d = 1$ m. several streamlines of the flow are plotted as shown in figure. Predict the magnitude of the velocity at the point A.

Analysis: The discharge rate per unit width between the bottom wall $(\psi_{wall} = 0)$ and the dividing streamline $(\psi_{dividing} = 3 \text{ m}^2/\text{s})$ is given by

$$\frac{Q}{d} = \psi_{dividing} - \psi_{wall} = (3-0) = 3 \text{ m}^2/\text{sec}$$

Entire sink flow go through the orifice. Since the water channel is 1 m wide, thus the total discharge is

$$Q = \frac{Q}{d}.d = (3).(1) = 3 \text{ m}^3/\text{sec}$$

To calculate the velocity at the point A, we must find out the distance (δ) between the two streamlines that enclose the point A. From the plotted figure, it is clearly observed that the streamline 1.8 is about 0.2 m away from streamline 1.6 which is close to the point A. the discharge rate per unit width between these two streamlines is equal to the difference in the value of the stream function. Thus, velocity of the flow at the point a is calculated as

$$v \cong \frac{Q}{d\delta} = \frac{1}{\delta} \cdot \frac{Q}{d} = \frac{1}{\delta} \cdot (\psi_{1.8} - \psi_{1.6})$$

$$\Rightarrow v = \frac{1}{0.2}(1.8 - 1.6) \; m^2/s = 0.93 \; m/s$$

2. DYNAMICS OF FLOWS

INTRODUCTION

In this chapter, we will discuss the various types of flows. We will learn how these flows behave by visualize their stream functions and velocity potential functions with respect to the reference planes. Dynamics of flows involves the study of fundamental parameters which governs the fluid flows. So, we will study the derivation of equation of continuity in different coordinates system which provides the values of any particular fluid's motion. We knew that a fluid flows continuously or in other words, we can say that a fluid flow is possible if its flow field satisfy the continuity equation. Thus, we consider equation of continuity as basic equation for fluid dynamism. As reader, we should keep in mind the equation of continuity is truly based on the Law of conservation of mass (which is a universal principle). In this chapter, this law is explained with illustrations.

UNIFORM FLOW

The fluid flow where its velocity remains constant during entire movement is considered as uniform flow. This can be further classified into two category such as

1. Flow which is parallel to $x-axis$.
2. Flow which is parallel to $y-axis$.

Uniform flow which is parallel to $x-axis$

INTRODUCTION TO FLUID KINEMATICS

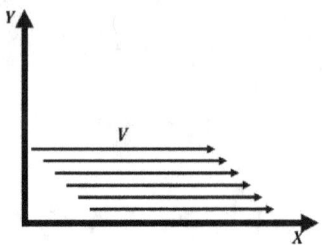

Uniform flow which is parallel to x – axis indicates that all fluid particles are moving in the direction which is parallel to x – axis as shown in figure. Let V be the fluid's motion which is constant along x – axis.

Assume that u and v are the components of fluid's motion V along x and y axis. Thus, we have,

$$u = V \text{ and } v = 0 \Rightarrow V = u\hat{i} + v\hat{j} = u\,\hat{i} \Rightarrow |V| = V = u$$

The stream function for the fluid' motion are defined as

$$u = \frac{\partial \psi}{\partial y} \text{ and } v = -\frac{\partial \psi}{\partial x}$$

The potential function for the fluid's motion are defined as

$$u = \frac{\partial \phi}{\partial x} \text{ and } v = \frac{\partial \phi}{\partial y}$$

Using above relations, the expression for fluid's motion rewritten as

$$V = u = \frac{\partial \psi}{\partial y} = \frac{\partial \phi}{\partial x}$$

Consider the expression

$$u = \frac{\partial \psi}{\partial y}$$

$$\Rightarrow d\psi = u\,dy$$

Integrating above equation w.r.t. y, we get

$$\Rightarrow \psi = u.y + K_1$$
$$\Rightarrow \psi = V.y + K_1$$

Where K_1 is constant of integration.

Next, consider the expression

$$u = \frac{\partial \phi}{\partial x}$$
$$\Rightarrow d\phi = u\,dx$$

Integrating above equation w.r.t. x, we get

$$\Rightarrow \phi = u.x + K_2$$
$$\Rightarrow \phi = V.x + K_2$$

Where K_2 is constant of integration.

Then let us plot the streamlines and potential lines for uniform flow parallel to x – axis.

The streamlines of Uniform flow is given by

$$\psi = V.y + K_1$$

Assume the initial condition as $\psi = 0$ when $y = 0$, putting these values in the above equation, we get

$$0 = V.0 + K_1 \Rightarrow K_1 = 0$$

Then equation of stream lines of Uniform flow becomes as

$$\psi = V.y \quad \text{-------(1)}$$

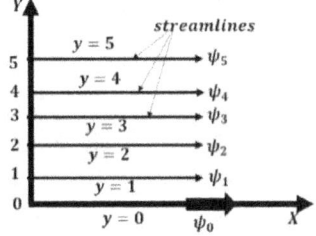 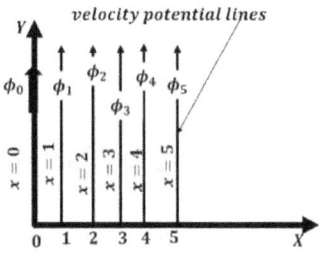

The streamlines are straight lines parallel to x – axis and at a distance y from the x – axis as shown in figure.

The potential lines for Uniform flow is given by

$$\phi = V.x + K_2$$

INTRODUCTION TO FLUID KINEMATICS

Assume the initial condition as $\phi = 0$ when $x = 0$, putting these values in the above equation, we obtain as $0 = V.0 + K_2 \Rightarrow K_2 = 0$. then equation of potential lines of Uniform flow becomes as

$\phi = V.x$ ------(2)

The potential lines are straight lines parallel to $y-$ axis and at a distance x from the $y-$ axis as shown in figure.

Uniform flow which is parallel to $y-axis$

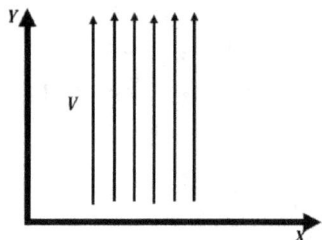

Uniform flow which is parallel to $y-$ axis indicates that all fluid particles are moving in the direction which is parallel to $y-$ axis as shown in figure. Let V be the fluid's motion which is constant along $y-$ axis.

Assume that u and v are the components of fluid's motion V along x and y axis. Thus, we have,

$$u = 0 \text{ and } v = V \Rightarrow V = u\hat{i} + v\hat{j} = v\hat{i} \Rightarrow |V| = V = v$$

The stream function for the fluid' motion are defined as

$$u = \frac{\partial \psi}{\partial y} \text{ and } v = -\frac{\partial \psi}{\partial x}$$

The potential function for the fluid's motion are defined as

$$u = \frac{\partial \phi}{\partial x} \text{ and } v = \frac{\partial \phi}{\partial y}$$

Using above relations, the expression for fluid's motion rewritten as

$$V = v = -\frac{\partial \psi}{\partial x} = \frac{\partial \phi}{\partial y}$$

INTRODUCTION TO FLUID KINEMATICS

Consider the expression

$$v = -\frac{\partial \psi}{\partial x}$$

$$\Rightarrow d\psi = -v\,dx$$

Integrating above equation w.r.t. x, we get

$$\Rightarrow \psi = -v.x + K_1 \Rightarrow \psi = -V.x + K_1$$

Where K_1 is constant of integration.

Next, consider the expression

$$u = \frac{\partial \phi}{\partial y} \Rightarrow d\phi = u\,dy$$

Integrating above equation w.r.t. y, we get

$$\Rightarrow \phi = u.y + K_2 \Rightarrow \phi = V.y + K_2$$

Where K_2 is constant of integration.

Then let us plot the streamlines and potential lines for uniform flow parallel to x – axis.

The streamlines of Uniform flow is given by

$$\psi = -V.x + K_1$$

Assume the initial condition as $\psi = 0$ when $x = 0$, putting these values in the above equation, we get

$$0 = V.0 + K_1 \Rightarrow K_1 = 0$$

Then equation of stream lines of Uniform flow becomes as

$$\psi = -V.x \quad \text{-------(1)}$$

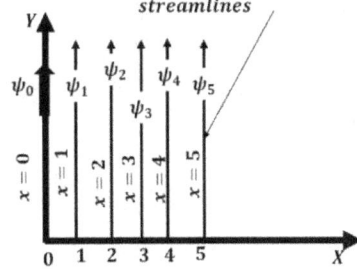
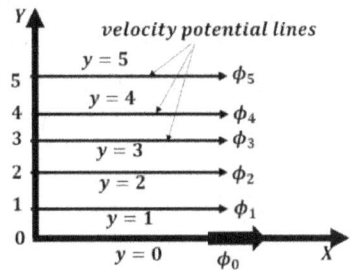

INTRODUCTION TO FLUID KINEMATICS

The streamlines are straight lines parallel to y – axis and at a distance x from the y – axis as shown in figure.

The potential lines for Uniform flow is given by

$$\phi = V.y + K_2$$

Assume the initial condition as $\phi = 0$ when $y = 0$, putting these values in the above equation, we obtain as $0 = V.0 + K_2 \Rightarrow K_2 = 0$. then equation of potential lines of Uniform flow becomes as

$$\phi = V.y \quad \text{------(2)}$$

The potential lines are straight lines parallel to x – axis and at a distance y from the x – axis as shown in figure.

SOURCE FLOW

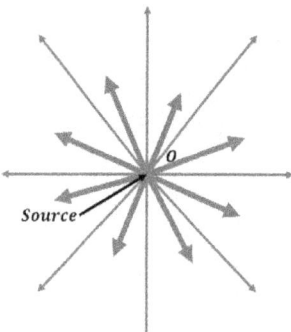

The source flow is the flow originated from a fixed point (considered as source) and tends to moving out radially in all directions of a plane with uniform fluid's motion as shown in figure. Let us find out the expression for this source flow as : $u_r = \dfrac{v}{2\pi r}$

and $u_\theta = 0$

Where u_r is the fluid motion in the radial direction at a radius r from the source O, u_θ is the component of fluid motion in tangential direction and v is the volume flow rate per unit depth. Since flow is originated from the source in radial direction only, thus tangential motion is zero. Let us derive the expression of stream function and velocity potential.

As per the definition of stream function in terms of radial and tangential components, we have

$$u_r = \frac{1}{r}\frac{\partial \psi}{\partial \theta} \quad \text{and} \quad u_\theta = -\frac{\partial \psi}{\partial r}$$

As per the definition of source flow, we have

$$u_r = \frac{v}{2\pi r} \text{ and } u_\theta = 0$$

Then we obtain as

$$u_r = \frac{1}{r}\frac{\partial \psi}{\partial \theta} = \frac{v}{2\pi r} \Rightarrow d\psi = r \cdot \frac{v}{2\pi r} d\theta \Rightarrow d\psi = \frac{v}{2\pi} d\theta$$

Integrating the above equation w.r.t. θ, we get

$$\psi = \frac{v}{2\pi} \times \theta + K_1$$

Assume the initial condition as $\psi = 0$ when $\theta = 0$, putting these values in the above equation, we get

$$0 = \frac{v}{2\pi} \times 0 + K_1 \Rightarrow K_1 = 0$$

Putting this value of K_1 in the above equation, we get the equation of stream function as

$$\psi = \frac{v}{2\pi} \cdot \theta$$

Where v is constant. The expression shows that the stream function is a function of θ. Thus, for each chosen value of θ, the stream function represent streamline in radial direction and a family of curves represents streamlines for the source flow field for different values of θ in radians as shown in figure.

INTRODUCTION TO FLUID KINEMATICS

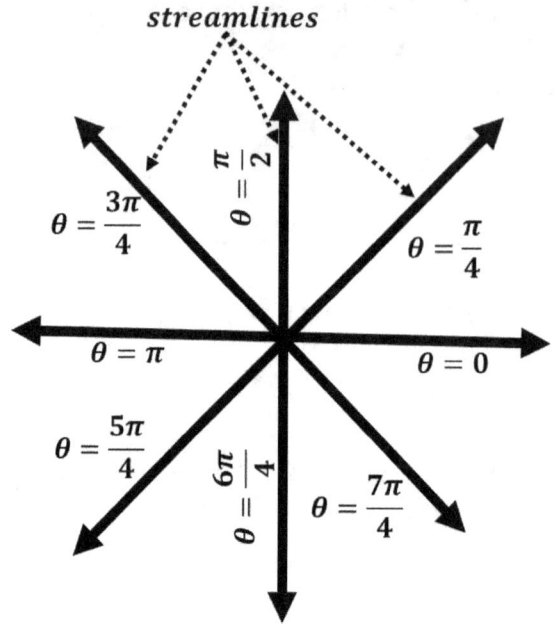

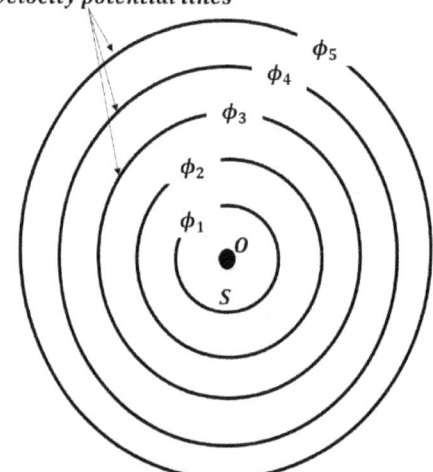

As per the definition of potential function in terms of radial and tangential components, we have

$$u_r = \frac{\partial \phi}{\partial r} \quad \text{and} \quad u_\theta = \frac{1}{r}\cdot\frac{\partial \phi}{\partial \theta}$$

As per the definition of source flow, we have

$$u_r = \frac{v}{2\pi r} \text{ and } u_\theta = 0$$

We have

$$u_r = \frac{\partial \phi}{\partial r} = \frac{v}{2\pi r} \Rightarrow d\phi = \frac{v}{2\pi r}.dr$$

Integrating the above equation, we get

$$\int d\phi = \int \frac{v}{2\pi r}.dr$$

$$\Rightarrow \phi = \frac{v}{2\pi} \int \frac{1}{r} dr$$

$$\Rightarrow \phi = \frac{v}{2\pi}.\log_e r$$

Where $\frac{v}{2\pi}$ is constant. The above equation indicates that the velocity potential function is constant for different values of r. Hence, graphical representation of velocity potential function of the source flow is a concentric circle with origin at the source as shown in figure.

SINK FLOW

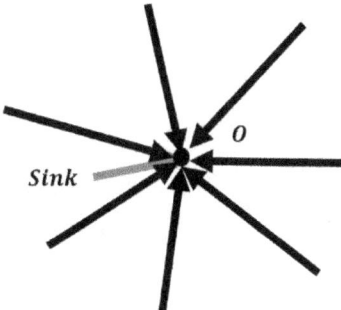

The sink flow is the flow originated from a fixed point (considered as source) and tends to moving inward in all directions of a plane with uniform fluid's motion as shown in figure. Let us find out the expression for this sink flow as : $u_r = \frac{v}{2\pi r}$ and $u_\theta = 0$

The graphical representation of streamlines and velocity potential lines of a sink flow is same as that of a source flow.

INTRODUCTION TO FLUID KINEMATICS

FREE-VORTEX FLOW

The circulatory fluid flow whose streamlines are concentric circles are known as **Free-vortex flow.** Mathematically, it can be express as $u_\theta \times r = K(\text{constant})$

The circulation around a stream line of a irrotational vortex is
$\Gamma = 2\pi r \times u_\theta = 2\pi \times K$

Where u_θ is tangential component of the fluid motion which is given by

$$u_\theta = \frac{\Gamma}{2\pi r}$$

Here, Γ is taken positive as free vortex is anticlockwise.

As per the definition of stream function in terms of radial and tangential components, we have

$$u_r = \frac{1}{r}\frac{\partial \psi}{\partial \theta} \quad \text{and} \quad u_\theta = -\frac{\partial \psi}{\partial r}$$

As per the definition of free vortex flow,

$$u_r = 0 \quad \text{and} \quad u_\theta = \frac{\Gamma}{2\pi r}$$

Thus, equating the values of u_θ, we get

$$u_\theta = -\frac{\partial \psi}{\partial r} = \frac{\Gamma}{2\pi r} \Rightarrow d\psi = -\frac{\Gamma}{2\pi r}.dr$$

Integrating the above expression, we obtain as

$$\int d\psi = \int -\frac{\Gamma}{2\pi r} dr = \left(-\frac{\Gamma}{2\pi}\right)\int \frac{1}{r}$$

$$\Rightarrow \psi = \left(-\frac{\Gamma}{2\pi}\right)\log_e r$$

Since $\frac{\Gamma}{2\pi}$ is constant, so the stream function is constant for the particular value of r. Hence the streamlines are concentric circles as shown in figure.

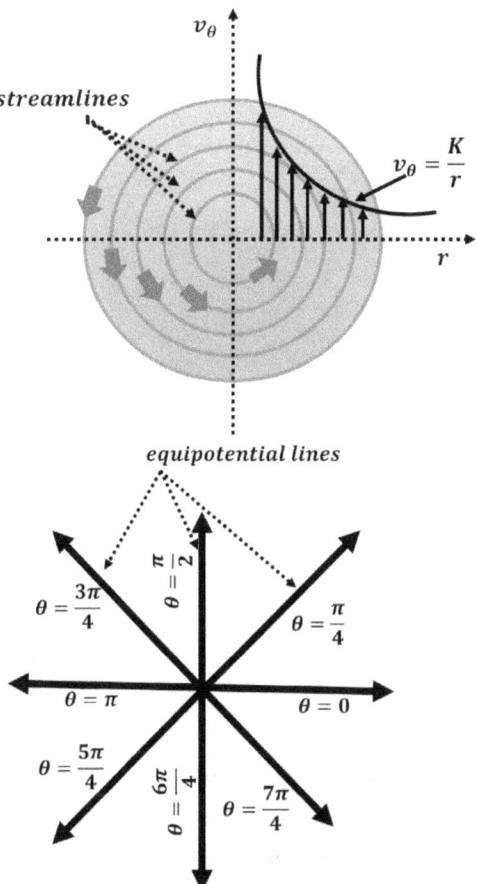

As per the definition of potential function in terms of radial and tangential components, we have

$$u_r = \frac{\partial \phi}{\partial r} \quad \text{and} \quad u_\theta = \frac{1}{r}\cdot\frac{\partial \phi}{\partial \theta}$$

As per the definition of free vortex flow, we have $u_r = 0$ and $u_\theta = \dfrac{\Gamma}{2\pi r}$

Thus, equating the values of u_θ, we get

INTRODUCTION TO FLUID KINEMATICS

$$\frac{1}{r}\frac{\partial \phi}{\partial \theta} = \frac{\Gamma}{2\pi r}$$

$$\Rightarrow d\phi = r.\frac{\Gamma}{2\pi r}.d\theta$$

$$\Rightarrow d\phi = \frac{\Gamma}{2\pi}.d\theta$$

Integrating the above expression, we obtain as

$$\int d\phi = \int \frac{\Gamma}{2\pi}$$

$$\Rightarrow \phi = \frac{\Gamma}{2\pi}\int d\theta$$

$$\Rightarrow \phi = \frac{\Gamma}{2\pi}.\theta$$

Since $\frac{\Gamma}{2\pi}$ is constant, so the potential function is constant for the particular value of θ. Thus, equipotential lines are radial as shown in figure.

PRINCIPLE OF CONSERVATION OF MASS

Statement: The resultant mass transfer during a time interval δt is equivalent to the total change in the whole mass during the flowing through any system during the same time interval δt. In other words, we can say

(mass enter into system during δt) − (mass exit from system during δt)
= (total change in whole mass during flow)

Mathematically, we can express the principle as

$$M_{in} - M_{out} = \delta M_{system}$$

$$\Rightarrow M_{in} - M_{out} = M_{final} - M_{initial}$$

This principle can apply on mass flow rate undergone through the system. It can be express as

$$\dot{M}_{in} - \dot{M}_{out} = \frac{\partial M_{system}}{\partial t}$$

INTRODUCTION TO FLUID KINEMATICS

$$\Rightarrow \frac{\partial M_{system}}{\partial t} - \left(\dot{M}_{in} - \dot{M}_{out}\right) = 0$$

$$\Rightarrow \frac{\partial M_{system}}{\partial t} + \dot{M}_{out} - \dot{M}_{in} = 0 \qquad \text{------(1)}$$

Consider an infinitesimal control volume. The mass of the given control volume is $dM = \rho dV$. The total mass at any instant in time t is given by $M = \int \rho dV$

Then rate of change of mass within the control volume is given by

$$\frac{dM}{dt} = \frac{d}{dt}\int \rho dV$$

On applying the principle of conservation of mass on control volume, we have

$$\frac{dM}{dt} = 0 \qquad \text{--------(2)}$$

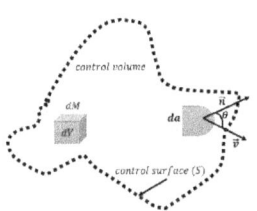

The above relation holds all types of control volumes whether it is fixed, moving, or deforming. Next, consider the mass flow entering and leaving the control volume through a differential area da of a fixed control volume. Let $\vec{n}$ be outward normal vector of da and $\vec{v}$ be the fluid motion relative to a fixed coordinate system as shown in figure. The mass flow rate is proportional to the normal component of fluid motion $\vec{v}\cos\theta$ ranging from a maximum outflow at $\theta = 0$ to a minimum of zero at $\theta = 90^0$ (flow is tangent to da) to a maximum inflow at $\theta = 180^0$ (flow is normal to da but in the opposite direction).

Since mass flow rate through cross-sectional area da is proportional to the fluid density (ρ), normal component of fluid motion $(v\cos\theta)$, and cross-sectional area (da) and can be express as

INTRODUCTION TO FLUID KINEMATICS

$$d\dot{M} = \rho(v\cos\theta)da = \rho(\vec{v}.\vec{n})da$$

The total mass flow rate entering and leaving from the control volume can be obtain by integrating $d\dot{M}$ over whole surface.

$$M_{out} - M_{in} = \int d\dot{M} = \int_S \rho(\vec{v}.\vec{n})da$$

$$\Rightarrow M_{out} - M_{in} = \int_S \rho v da \quad \text{-------(3)}$$

Assuming control surface is normal to the flow at all directions where it crosses the fluid flow. So, the dot product $\vec{v}.\vec{n}$ reduced to magnitude of the fluid motion and then $\rho(\vec{v}.\vec{n})da$ reduced to $\rho.v.da$ as mentioned in the equation (3). Now, splitting the surface integral in equation (3) into two parts i.e., one for the outgoing flow and other one for incoming flow, then expression reduced to

$$\dot{M}_{out} - \dot{M}_{in} = \int_S \rho v da = \sum_{out} \rho v da - \sum_{in} \rho v da$$

According to the principle of conservation of mass within the entire control volume, we have

$$\frac{dM}{dt} + \dot{M}_{out} - \dot{M}_{in} = 0$$

$$\Rightarrow \frac{d}{dt}\int \rho dV + \sum_{out} \rho v da - \sum_{in} \rho v da = 0$$

$$\Rightarrow \frac{d}{dt}\int \rho dV = \sum_{in} \rho v da - \sum_{out} \rho v da$$

EQUATION OF CONTINUITY

The equation based on the principle of conservation of mass is known as **equation of continuity.** Thus, equation of continuity indicates that the amount of discharge of a fluid flowing per second through a cross-sectional of a pipe or a channel is remains constant.

INTRODUCTION TO FLUID KINEMATICS

Let us derive the expression for the equation of continuity. Consider the two cross-sectional area of a pipe as shown in figure. Assume the following parameters as:

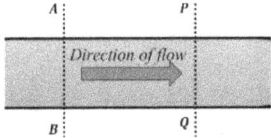

ρ_i = Density of a fluid flowing through the section AB

a_i = cross-sectional area of section AB

v_i = fluid motion at the section AB

ρ_d = Density of a fluid flowing through the section PQ

a_d = cross-sectional area of section PQ

v_d = fluid motion at the section PQ

The rate of fluid flow through the section AB = $\rho_i a_i v_i$

And The rate of fluid flow through the section PQ = $\rho_d a_d v_d$

According to the principle of conservation of mass, we have

The rate of fluid flow through the section AB = The rate of fluid flow through the section PQ

$$\Rightarrow \rho_i a_i v_i = \rho_d a_d v_d$$

Note: For an incompressible fluid, we have $\rho_i = \rho_d$ then, equation of continuity becomes

$$a_i v_i = a_d v_d$$

DERIVATION USING THE DIVERGENCE THEOREM

INTRODUCTION TO FLUID KINEMATICS

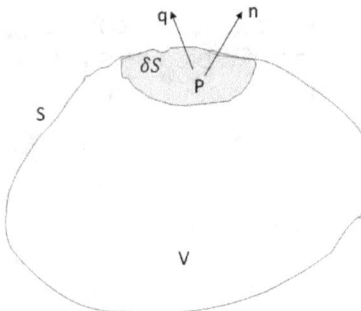

Let S be an arbitrary small closed surface drawn in the compressible fluid enclosing a volume V and let S be taken fixed in space. Let P (x, y, z) be any point S and let $\rho(x, y, z, t)$ be the fluid density at P at any time t. Let δS denote element of the surface S enclosing P. Let n be the unit outward-drawn normal at δS and let v be the fluid velocity at P. Then the normal component of v measured outwards from V is $n.v$

Thus, rate of mass flow across $\delta S = \rho\, (n.v)\delta S$

Thus, total rate of mass flow across S

$$= \int \rho\,(n.v)dS$$
$$= \int \nabla.(\rho v)dV \quad (1) \quad (by\ Gauss\ Divergence\ Theorem)$$

Total rate of mass flow into V $= -\int \nabla.(\rho v)dV$

Again, the mass of the fluid within S at time $t = \int \rho dV$

Thus, total rate of mass increase within S

$$= \frac{\partial}{\partial t}\int \rho dV = \int \frac{\partial \rho}{\partial t}dV \quad (2)$$

Suppose that the region V of the fluid contains neither sources nor sinks (i.e., there are no inlets or outlets through which fluid can enter or leave the region). Then by the law of conservation of the fluid mass, the rate of increase of the mass of fluid within V must be equal to the total rate of mass flowing into V. Hence from (1) and (2), we have

$$\int \frac{\partial \rho}{\partial t}dV = -\int \nabla.(\rho v)dV$$

$$\int \left[\frac{\partial \rho}{\partial t} + \nabla.(\rho v)\right]dV = 0$$

Which holds for arbitrary small volumes V.

INTRODUCTION TO FLUID KINEMATICS

$$\frac{\partial \rho}{\partial t} + \nabla \cdot (\rho v) = 0$$

Above equation is called *Equation of continuity*

EQUATION OF CONTINUITY FOR AN INFINITESIMAL CONTROL VOLUME

Consider an infinitesimal cuboid shaped control volume aligned with the axes in cartesian coordinate as shown in figure. The dimension of the box are $dx, dy,$ and dz, point K chosen arbitrarily located at the centre of cuboidal control surface. Let $u, v,$ and w are the components of the fluid motion which is flowing with density ρ. Using Taylor series expansion, the value of ρu at the point positioned at the center of the right face in $x-$ direction is given by

$$\rho u = \rho u + \frac{\partial (\rho u)}{\partial x}\frac{dx}{2} + \frac{1}{2!}\frac{\partial^2 (\rho u)}{\partial x^2}\left(\frac{dx}{2}\right)^2 + \ldots\ldots$$

Since the cuboidal control surface shrinks to a point results into the neglecting second-order and higher order terms are relevant. In fact, smaller dx, assumption of negligence of second order terms is more prominent. Applying truncated Taylor series expansion to the density times normal component of fluid motion at the center point of each of the six faces of the cuboid are given by

INTRODUCTION TO FLUID KINEMATICS

Center of right face: $\rho u \equiv \rho u + \dfrac{\partial(\rho u)}{\partial x}\dfrac{dx}{2}$

Center of left face: $\rho u \equiv \rho u - \dfrac{\partial(\rho u)}{\partial x}\dfrac{dx}{2}$

Center of front face: $\rho w \equiv \rho w + \dfrac{\partial(\rho w)}{\partial z}\dfrac{dz}{2}$

Center of rear face: $\rho w \equiv \rho w - \dfrac{\partial(\rho w)}{\partial z}\dfrac{dz}{2}$

Center of top face: $\rho v \equiv \rho v + \dfrac{\partial(\rho v)}{\partial y}\dfrac{dy}{2}$

Center of bottom face: $\rho v \equiv \rho v - \dfrac{\partial(\rho v)}{\partial y}\dfrac{dy}{2}$

From the figure, it can be observed that total mass flow rate entering the cuboidal control surface is given by

$$\sum_{in}\dot{M} = \underbrace{\left(\rho u - \dfrac{\partial(\rho u)}{\partial x}\dfrac{dx}{2}\right)dydz}_{\text{left face}}$$

$$+ \underbrace{\left(\rho v - \dfrac{\partial(\rho v)}{\partial y}\dfrac{dy}{2}\right)dxdz}_{\text{bottom face}} + \underbrace{\left(\rho w - \dfrac{\partial(\rho w)}{\partial z}\dfrac{dz}{2}\right)dxdy}_{\text{rear face}}$$

Then, total mass flow rate out from the cuboidal control surface is given by

$$\sum_{out}\dot{M} = \underbrace{\left(\rho u - \dfrac{\partial(\rho u)}{\partial x}\dfrac{dx}{2}\right)dydz}_{\text{right face}}$$

$$+ \underbrace{\left(\rho v + \dfrac{\partial(\rho v)}{\partial y}\dfrac{dy}{2}\right)dxdz}_{\text{top face}} + \underbrace{\left(\rho w + \dfrac{\partial(\rho w)}{\partial z}\dfrac{dz}{2}\right)dxdy}_{\text{front face}}$$

INTRODUCTION TO FLUID KINEMATICS

Let us assume the cuboidal control volume shrinking to infinitesimal size, with dimensions dx, dy, and dz as shown in figure such that the entire control volume dissolve into a point in the given fluid flow field. Thus, rate of change of mass flow within the control volume is given by

$$\int_S \frac{\partial \rho}{\partial t} dV \equiv \frac{\partial \rho}{\partial t} dxdydz \quad \text{---------(1)}$$

The total mass flow rate entering and leaving from the control volume is given by

$$\sum_{in} \dot{M} - \sum_{out} \dot{M} = -\frac{\partial(\rho u)}{\partial x} dxdydz - \frac{\partial(\rho v)}{\partial y} dxdydz - \frac{\partial(\rho w)}{\partial z} dxdydz \quad \text{----(2)}$$

Applying the principle of conservation of mass on the entire control volume along with using equation (1) and (2), we have

$$\frac{\partial \rho}{\partial t} dxdydz = -\frac{\partial(\rho u)}{\partial x} dxdydz - \frac{\partial(\rho v)}{\partial y} dxdydz - \frac{\partial(\rho w)}{\partial z} dxdydz$$

Canceling the volume of control volume i.e., $dxdydz$, from the both side of above equation, we obtain as

$$\frac{\partial \rho}{\partial t} + \frac{\partial(\rho u)}{\partial x} + \frac{\partial(\rho v)}{\partial y} + \frac{\partial(\rho w)}{\partial z} = 0$$

The above equation for a compressible fluid is known as **Equation of Continuity**.

EQUATION OF CONTINUITY IN CARTESIAN COORDINATES

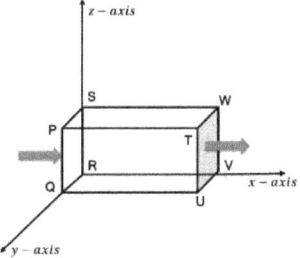

Let us assume a fluid element of cuboidal shape whose dimension are $\delta x, \delta y$, and δz w.r.t to three dimensional coordinate system. Let u, v, and w are the components of fluid motion in x, y, and z directions

INTRODUCTION TO FLUID KINEMATICS

respectively. Then we have,

Mass flow rate entering the face $PQRS$ per second is

= density × fluid motion in $x-$ axis
  × cross-sectional area of face PQRS

$= \rho \times u \times (\delta y . \delta z)$

$= \rho u \delta y \delta z$

Mass flow rate leaving the face $TUVW$ per second is

= entered mass
+rate of change of mass flow through face TUVW

$= \rho u \delta y \delta z + \dfrac{\partial}{\partial x}(\rho u \delta y \delta z) \delta x$

Thus, total mass flow rate occurs in $x-$ axis is given by

= mass flow through face PQRS − mass flow through TUVW

$= \rho u \delta y \delta z - \rho u \delta y \delta z - \dfrac{\partial}{\partial x}(\rho u \delta y \delta z) \delta x$

$= -\dfrac{\partial}{\partial x}(\rho u \delta y \delta z) \delta x$

$= -\dfrac{\partial}{\partial x}(\rho u) \delta x \delta y \delta z \qquad [\because \ \delta y \delta z \text{ is constant}]$

Similarly, total mass flow rate occurs in $y-$ axis is given by

$= -\dfrac{\partial}{\partial y}(\rho v) \delta x \delta y \delta z$

Again, total mass flow rate occurs in $z-$ axis is given by

$= -\dfrac{\partial}{\partial z}(\rho w) \delta x \delta y \delta z$

Thus, mass flow rate within the entire cuboidal fluid element is given by

INTRODUCTION TO FLUID KINEMATICS

$$= -\left[\frac{\partial}{\partial x}(\rho u) + \frac{\partial}{\partial y}(\rho v) + \frac{\partial}{\partial z}(\rho w)\right]\delta x \delta y \delta z$$

We know that mass of the fluid preserves within a fluid element at instant time is given by

$$= \text{density} \times \text{volume}$$
$$= \rho \delta x \delta y \delta z$$

Then total rate of change in mass flow within cuboidal fluid element is given by

$$= \frac{\partial}{\partial t}(\rho \delta x \delta y \delta z)$$
$$= \frac{\partial \rho}{\partial t}\delta x \delta y \delta z$$

According to principle of mass conservation which states that the mass flow rate within particular fluid flow field is neither created nor destroyed results into the total mass flow rate within entire fluid element is equivalent to the total change in mass of the fluid in the fluid element during time interval t. So, we can write,

$$= -\left[\frac{\partial}{\partial x}(\rho u) + \frac{\partial}{\partial y}(\rho v) + \frac{\partial}{\partial z}(\rho w)\right]\delta x \delta y \delta z = \frac{\partial \rho}{\partial t}\delta x \delta y \delta z$$

$$\Rightarrow -\left[\frac{\partial}{\partial x}(\rho u) + \frac{\partial}{\partial y}(\rho v) + \frac{\partial}{\partial z}(\rho w)\right] = \frac{\partial \rho}{\partial t}$$

$$\Rightarrow \frac{\partial \rho}{\partial t} + \left[\frac{\partial}{\partial x}(\rho u) + \frac{\partial}{\partial y}(\rho v) + \frac{\partial}{\partial z}(\rho w)\right] = 0$$

This equation is known as **Equation of Continuity** in cartesian coordinates system.

INTRODUCTION TO FLUID KINEMATICS

EQUATION OF CONTINUITY IN CYLINDRICAL COORDINATES

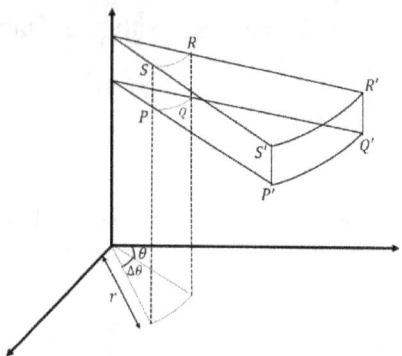

Assume that there is a fluid element occupy the space at point P whose cylindrical coordinates are (r, θ, z) where $r \geq 0, 0 \leq \theta \leq 2\pi, -\infty \leq z \leq \infty$. Let v and ρ are the motion and density of the fluid at the instant of time t. Thus, v_r, v_θ and v_z are the components of fluid motion acts on the faces PP', PQ and QR of the fluid element respectively. Let us assume the given fluid element is of cuboidal shape whose dimension are $\Delta r, r\Delta\theta$ and Δz w.r.t to three dimensional coordinate system respectively.

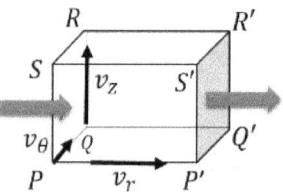

Mass flow rate entering the face $PQRS$ is

= density × component of fluid motion in PP' direction × surface area of face PQRS

$= \rho \times v_r \times (r\Delta\theta\Delta z) = \rho v_r r\Delta\theta\Delta z$

Mass flow rate entering the face $P'Q'R'S'$ is

= density × component of fluid motion in PP' direction × cross-sectional area of face P'Q'R'S'.

INTRODUCTION TO FLUID KINEMATICS

$$= \rho v_r r \Delta\theta \Delta z + \frac{\partial}{\partial r}(\rho v_r r \Delta\theta \Delta z)\Delta r$$

Thus, total mass flow rate occurs in $x-axis$ is given by

= mass flow through face PQRS − mass flow through P'Q'R'S'

$$= \rho v_r r \Delta\theta\Delta z - \rho v_r r \Delta\theta\Delta z - \frac{\partial}{\partial r}(\rho v_r r \Delta\theta\Delta z)\Delta r$$

$$= -\frac{\partial}{\partial r}(\rho v_r r \Delta\theta\Delta z)\Delta r \qquad \text{---------(1)}$$

$$= -\frac{\partial}{\partial r}(\rho v_r)\Delta r \Delta\theta\Delta z$$

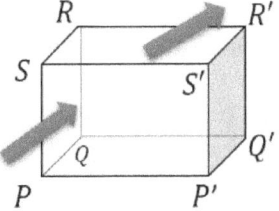

Similarly, the total mass flow rate occurs in $y-axis$ is given by

= mass flow through face PP'S'S − mass flow through face QQR'R

$$= \rho v_\theta \Delta r \Delta z - \rho v_\theta \Delta r \Delta z - \frac{\partial}{\partial \theta}(\rho v_\theta \Delta r \Delta z)\Delta\theta$$

$$= -\frac{\partial}{\partial \theta}(\rho v_\theta \Delta r \Delta z)\Delta\theta$$

$$= -\frac{\partial}{\partial \theta}(\rho v_\theta)\Delta r \Delta\theta\Delta z \qquad \text{---------(2)}$$

Again, the total mass flow rate occurs in $z-axis$ is given by

= mass flow through face QPPQ' − mass flow through RSS'R

INTRODUCTION TO FLUID KINEMATICS

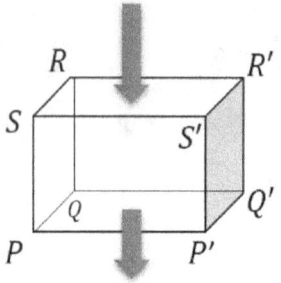

$$= \rho v_z r \Delta\theta\Delta r - \rho v_z r \Delta\theta\Delta r - \frac{\partial}{\partial z}(\rho v_z r \Delta\theta\Delta r)\Delta z$$

$$= -\frac{\partial}{\partial z}(\rho v_z r \Delta\theta\Delta r)\Delta z$$

$$= -\frac{\partial}{\partial z}(\rho v_z) r \Delta r \Delta\theta \Delta z \qquad \text{--------(3)}$$

Thus, mass flow rate within the entire cuboidal fluid element is given by

$$= -\left[\frac{\partial}{\partial r}(\rho r v_r) + \frac{\partial}{\partial \theta}(\rho v_\theta) + r\frac{\partial}{\partial z}(\rho v_z)\right]\Delta r \Delta\theta\Delta z \quad \text{------(4)}$$

We know that mass of the fluid preserves within a fluid element at instant time is given by

$$= \text{density} \times \text{volume}$$
$$= \rho r \Delta r \Delta\theta \Delta z$$

Then total rate of change in mass flow within cuboidal fluid element is given by

$$= \frac{\partial}{\partial t}(\rho r \Delta r \Delta\theta \Delta z)$$

$$= \frac{\partial \rho}{\partial t} r \Delta r \Delta\theta \Delta z \qquad \text{--------(5)}$$

According to principle of mass conservation which states that the mass flow rate within particular fluid flow field is neither created nor destroyed results into the total mass flow rate within entire fluid element is equivalent to the total change in mass of the fluid in the fluid element at time t. So, using equations (4) and (5); we can write,

$$r\Delta r\Delta\theta\Delta z\frac{\partial\rho}{\partial t}=-\left[\frac{\partial}{\partial r}(\rho rv_r)+\frac{\partial}{\partial\theta}(\rho v_\theta)+r\frac{\partial}{\partial z}(\rho v_z)\right]\Delta r\Delta\theta\Delta z$$

$$\Rightarrow \frac{\partial\rho}{\partial t}+\left[\frac{1}{r}\frac{\partial}{\partial r}(\rho rv_r)+\frac{1}{r}\frac{\partial}{\partial\theta}(\rho v_\theta)+\frac{\partial}{\partial z}(\rho v_z)\right]=0$$

The above equation is known as **Equation of continuity in cylindrical coordinates** for an unsteady, incompressible fluid flow.

EQUATION OF CONTINUITY IN POLAR COORDINATES

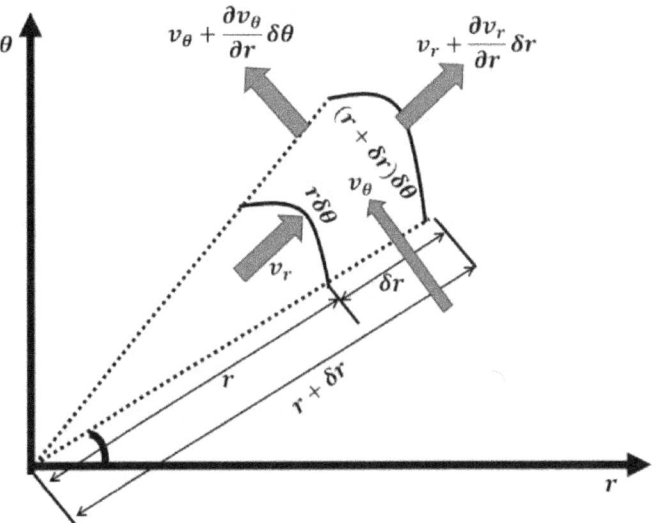

Let us assume that a fluid element of a two-dimensional incompressible flow positioned at the point O whose polar coordinates are (r,θ) where $r\geq 0, 0\leq\theta\leq 2\pi$. Let v and ρ are the motion and density of the fluid at the instant of time t. Then, v_r and v_θ are the components of fluid motion in radial and angular direction respectively. Consider the fluid element makes an angle θ with the reference axis and $\delta\theta$ be the increment in angle due to movement of fluid element in particular flow field. The dimension of the fluid element are

INTRODUCTION TO FLUID KINEMATICS

$PQ = r\delta\theta, QR = \delta r, RS = (r+\delta r)\delta\theta, SP = \delta r$ as shown in figure.

<u>Consider the flow in radial direction</u>

Mass flow rate entering the face PQ is

$= \rho \times$ fluid motion in r-direction $\times$ area

$= \rho \times v_r \times (r\delta\theta \times 1)$ [$\because$ thickness of fluid element $= 1$]

$= \rho v_r r\delta\theta$

Mass flow rate leaving the face RS is

$= \rho \times$ fluid motion in r-direction $\times$ area

$= \rho \times \left(v_r + \dfrac{\partial v_r}{\partial r}\delta r\right) \times (RS \times 1)$

$= \rho \times \left(v_r + \dfrac{\partial v_r}{\partial r}\delta r\right) \times (r+\delta r)\delta\theta$

$= \rho \left(v_r.r + v_r.\delta r + r.\dfrac{\partial v_r}{\partial r}\delta r\right)\delta\theta$

$\left[\because (\delta r)^2 \text{ is very small and thus, neglected}\right]$

Total mass flow rate in $r - direction$ is equal to

= mass flow through face PQ $-$ mass flow through face RS

$= \rho.v_r.r\delta\theta - \rho\left(v_r.r + v_r\delta r + r.\dfrac{\partial v_r}{\partial r}.\delta r\right)\delta\theta$

$= -\rho\left(v_r.\delta r + r\dfrac{\partial v_r}{\partial r}.\delta r\right)\delta\theta$

$= -\rho\left(\dfrac{v_r}{r} + \dfrac{\partial v_r}{\partial r}\right)r.\delta r.\delta\theta$

Next, consider the flow in $\theta - direction$ is equal to

Mass flow rate entering the face QR is

INTRODUCTION TO FLUID KINEMATICS

$= \rho \times$ fluid direction in θ – direction $\times$ area

$= \rho \times v_\theta \times (\delta r.1)$

$= \rho v_\theta \delta r$

Mass flow rate leaving the face RP is

$= \rho \times$ fluid motion in θ – direction $\times$ area

$= \rho \times \left(v_\theta + \dfrac{\partial v_\theta}{\partial \theta} . \delta\theta \right) \times (\delta r.1)$

$= \rho \left(v_\theta + \dfrac{\partial v_\theta}{\partial \theta} . \delta\theta \right) . \delta r$

Total mass flow rate in θ – direction is equal to mass flow rate through face QR – mass flow rate through face RP

$= \rho . v_\theta . \delta r - \rho \left(v_\theta + \dfrac{\partial v_\theta}{\partial \theta} \delta\theta \right) \delta r$

$= -\rho \left(\dfrac{\partial v_\theta}{\partial \theta} \delta\theta \right) \delta r$

$= -\rho . \dfrac{\partial v_\theta}{\partial \theta} . \dfrac{r.\delta\theta \delta r}{r}$

Moreover, mass flow within flow field
$= \rho \times$ volume of fluid element

$= \rho \times (r\delta\theta . \delta r . 1)$

$= \rho . r\delta\theta . \delta r$

Mass flow rate within the fluid element at instant time t

$= \dfrac{\partial}{\partial t} (\rho . r\delta\theta . \delta r)$

$= \dfrac{\partial \rho}{\partial t} . r\delta\theta . \delta r \qquad \left[\because \text{fluid is unsteady} \Rightarrow \dfrac{\partial \rho}{\partial t} \neq 0 \right]$

According to principle of mass conservation which states

INTRODUCTION TO FLUID KINEMATICS

that the mass flow rate within particular fluid flow field is neither created nor destroyed results into the total mass flow rate within entire fluid element is equivalent to the total change in mass of the fluid in the fluid element at time t. So, we can write,

$$-\rho\left(\frac{v_r}{r}+\frac{\partial v_r}{\partial r}\right)r\delta r\delta\theta - \rho\frac{\partial v_\theta}{\partial \theta}\cdot\frac{r\delta\theta\delta r}{r} = \frac{\partial \rho}{\partial t}r\delta r\delta\theta$$

$$\Rightarrow -\rho\left(\frac{v_r}{r}+\frac{\partial v_r}{\partial r}\right) - \rho\frac{\partial v_\theta}{\partial \theta}\cdot\frac{1}{r} = \frac{\partial \rho}{\partial t}$$

$$\Rightarrow \frac{\partial \rho}{\partial t} + \rho\left(\frac{v_r}{r}+\frac{\partial v_r}{\partial r}\right) + \rho\frac{\partial v_\theta}{\partial \theta}\cdot\frac{1}{r} = 0$$

This equation is known as the equation of continuity in polar coordinates for two-dimensional unsteady flow.

DEDUCTIONS

DEDUCTION 1 Recall the Equation of continuity in polar coordinates for two-dimensional unsteady flow as

$$\frac{\partial \rho}{\partial t} + \rho\left(\frac{v_r}{r}+\frac{\partial v_r}{\partial r}\right) + \rho\frac{\partial v_\theta}{\partial \theta}\cdot\frac{1}{r} = 0$$

If fluid flow is steady, then we have $\frac{\partial \rho}{\partial t} = 0$. In this case, Equation of continuity becomes

$$\rho\left(\frac{v_r}{r}+\frac{\partial v_r}{\partial r}\right) + \rho\frac{\partial v_\theta}{\partial \theta}\cdot\frac{1}{r} = 0$$

$$\Rightarrow \frac{v_r}{r}+\frac{\partial v_r}{\partial r}+\frac{\partial v_\theta}{\partial \theta}\cdot\frac{1}{r} = 0$$

$$\Rightarrow v_r + r\cdot\frac{\partial v_r}{\partial r}+\frac{\partial v_\theta}{\partial \theta} = 0$$

$$\Rightarrow \frac{\partial}{\partial r}(r.v_r) + \frac{\partial v_\theta}{\partial \theta} = 0$$

DEDUCTION 2 Recall the Equation of continuity in cartesian coordinates for three-dimensional unsteady flow as $\dfrac{\partial \rho}{\partial t}+\left[\dfrac{\partial}{\partial x}(\rho u)+\dfrac{\partial}{\partial y}(\rho v)+\dfrac{\partial}{\partial z}(\rho w)\right]=0$

If fluid flow is steady, then we have $\dfrac{\partial \rho}{\partial t}=0$. In this case, Equation of continuity becomes

$$\dfrac{\partial}{\partial x}(\rho u)+\dfrac{\partial}{\partial y}(\rho v)+\dfrac{\partial}{\partial z}(\rho w)=0$$

DEDUCTION 3   Recall the Equation of continuity in cartesian coordinates for three-dimensional unsteady flow as $\dfrac{\partial \rho}{\partial t}+\left[\dfrac{\partial}{\partial x}(\rho u)+\dfrac{\partial}{\partial y}(\rho v)+\dfrac{\partial}{\partial z}(\rho w)\right]=0$

If fluid flow is steady, then we have $\dfrac{\partial \rho}{\partial t}=0$.

If fluid is incompressible, then we have $\rho=$ constant. In this case, Equation of continuity becomes

$$\rho\left[\dfrac{\partial}{\partial x}u+\dfrac{\partial}{\partial y}v+\dfrac{\partial}{\partial z}w\right]=0$$

$$\Rightarrow \dfrac{\partial}{\partial x}u+\dfrac{\partial}{\partial y}v+\dfrac{\partial}{\partial z}w=0$$

DEDUCTION 4 The Equation of continuity in spherical polar coordinates for three-dimensional unsteady flow as

$$\dfrac{\partial \rho}{\partial t}+\dfrac{1}{r^2}\dfrac{\partial}{\partial r}\left(\rho r^2 v_r\right)+\dfrac{1}{r\sin\theta}\dfrac{\partial}{\partial \theta}\left(\rho \sin\theta v_\theta\right)+\dfrac{1}{r\sin\theta}\dfrac{\partial}{\partial \phi}\left(\rho v_\phi\right)=0$$

DEDUCTION 5   The Equation of continuity derive from Divergence theorem is

$$\dfrac{\partial \rho}{\partial t}+\nabla.(\rho.V)=0$$

Where ρ and V are density and fluid motion of an incompressible flow which is free from sources and sinks.

INTRODUCTION TO FLUID KINEMATICS

Since the term $\nabla.(\rho.V)$ rewritten as

$\nabla.(\rho.V) = \rho.\nabla.V + \nabla\rho.V$,

then Equation of continuity becomes

$$\frac{\partial \rho}{\partial t} + \rho.\nabla V + \nabla \rho.V = 0$$

$$\frac{\partial \rho}{\partial t} + \rho.\nabla V = 0 \quad [\because \nabla \rho = 0]$$

If fluid flow is steady, then we have $\frac{\partial \rho}{\partial t} = 0$. In this case, Equation of continuity becomes

$$\rho.\nabla V = 0$$
$$\Rightarrow \nabla V = 0$$
$$\Rightarrow \frac{\partial}{\partial x}u + \frac{\partial}{\partial y}v + \frac{\partial}{\partial z}w = 0$$

BOOSTER CAPSULE: NO SLIP CONDITION

We has been observed that water in river flows over large rocks. Such example illustrate the flow of a fluid over the solid surfaces (considered as non-porous) where it is essential to know that whether presence of solid surfaces does affect the fluid flow or not ?. It has been observed that speed of water flow perpendicular to solid surfaces are close to zero and speed of water flow closely over the sold surfaces comes to complete stop. The similar observations are proven from all the experiments conducted to analyse the nature of fluid flows over the solid surfaces. Scientifical data reveals that speed of a fluid comes to almost zero at the close interface between the solid surface and fluid layers. In this condition, a fluid is in direct contact with the solid surface but at same instant, fluid flow acquire zero velocity relative to the solid surface. This condition is well known as " **no-slip condition**".

INTRODUCTION TO FLUID KINEMATICS

IMPORTANCE OF "NO-SLIP CONDITION" IN THE STUDY OF FLUID MECHANICS

1. VISCOSITY

Viscosity of a fluid develops due to frictional forces acts on different layers of fluids. The viscosity of liquids is due to cohesive forces between the molecules in liquids, also viscosity of gases is due to the gaseous molecular collisions. The viscosity would be vary for the different fluids. For example, for Newtonian fluids, the relationship between viscosity and rate of deformation is linear due to which it vary strongly with temperature. While reverse observation shown by non-Newtonian fluids where the relationship between viscosity and rate of deformation is not linear.

When two fluids comes in contact with each other, then an interface boundary conditions occurs at the free surface of a fluid. We illustrate a simple example where one fluid is water and another fluid is air. Then interface is the flat free surface of water (imagine that water is flowing in a calm river). Boundary conditions at water-air interface is

$$v_{water} = v_{air} \quad \tau_{water} = \mu_{water}\left(\frac{du}{dy}\right)_{water} \quad \tau_{air} = \mu_{air}\left(\frac{du}{dy}\right)_{air}$$

From the fluid property table, data evident that kinematic viscosity of water μ_{water} is 50 times greater than kinematic viscosity of air μ_{air}. This indicates that flowing water moves smoothly and flowing air induces very less amount of resistance force to the flowing water as a result the air does not slow down the speed of water. Here, no-slip condition occurs at the free surface of water. In such flows the water is no "slip between the fluid (air) and the free surface of water.

2. **BOUNDARY LAYER**

To analyse the fluid flow, we need to assign proper frame of reference and boundary conditions. The most commonly used boundary condition is the **no-slip condition,** which infer that there is no slip between the fluid and the solid surface. In other words, it indicates that the velocity of fluid particles adhere to the solid surface is almost zero which is equivalent to the velocity of a stationary solid surface. Consider an example of the thin film of water between a piston and its spherical wall. A degenerate form of no slip condition occurs at the thin film of fluid (water) where the velocity of fluid particles moving with piston is almost zero, but the velocity of fluid particles adhere to the spherical wall is $\vec{v}_{surface} = -v_{piston}\vec{j}$.

Thus, specifications of geometry, application of no-slip conditions, integration of differential equations are few steps which should proceeds to analyse any particular flow problem.

3. **FLOW SEPARATION**

Consider the flow of a fluid over a curved surface where the no-slip condition applies along the entire surface, even downstream of the flow. Since flow separation occurs at separation point where boundary layer does not form on the free surface and deviates from the free surface. The no-slip condition applies at flow separation point because measurements of quantities derived from the velocity field where it has assumed that tangential velocity component is zero at a stationary solid surface.

4. **IDENTIFICATION OF FLUID FLOW**

Various study proposed that no-slip condition applicable to those fluids where boundary conditions holds. For example, Newtonian fluid flows over solid boundary where fluid velocity at solid surface is proportional to the fluid shear stress at the free surface and covers slip length. However, no-slip condition is not applicable to nanofluids where continuum hypothesis does not hold. Also, fluids flow over the free surface having nanometre dimension unable to hold no-slip condition. Couette flow is the fluid flow over infinite plates and Poiseuille flow is the fluid flow through confined boundaries. The no-slip condition verified the fluid flow at the boundaries. Thus, it is necessary to check no-slip conditions along with other boundary conditions to identify the types of fluid flow.

5. **SKIN FRICTION DRAG**

A unsteady fluid flow where a liquid (usually water) and a gas (usually air) generates boundary conditions at the free surface of liquid. In this case, water is flows horizontally and surface tension effects are insignificant. The shear stress acting on the fluid particles of the water is equal to shear stress acting on the fluid particles of the air just above the surface.

$$\tau_{water} = \tau_{air}$$

$$\Rightarrow \mu_{water} \left[\frac{du}{dy}\right]_{water} = \mu_{air} \left[\frac{du}{dy}\right]_{air}$$

From the fluid property tables, it can conclude that dynamic viscosity of water i.e., μ_{water} is 50 times greater than dynamic viscosity of air i.e., μ_{air}. In order to equalize the above equation, velocity gradient i.e., $\left[\frac{du}{dy}\right]_{air}$ should be 50 times greater than $\left[\frac{du}{dy}\right]_{water}$. It implies that flowing water drags the air along with it. The movement of fluid particles of water

INTRODUCTION TO FLUID KINEMATICS

is slow down only in boundary layer. Due to the presence of no-slip condition at the boundary layer reduce the frictional drag acts on the fluid particles.

HIGHER ORDER THINKING SKILL QUESTION

Q 1. What is Doublet in uniform flow?

Solution: Doublet is a combination of source flow and sink flow such a way that both of them are approaching each other so that the distance between then tends to zero and doublet strength $2d.k$ remains constant.

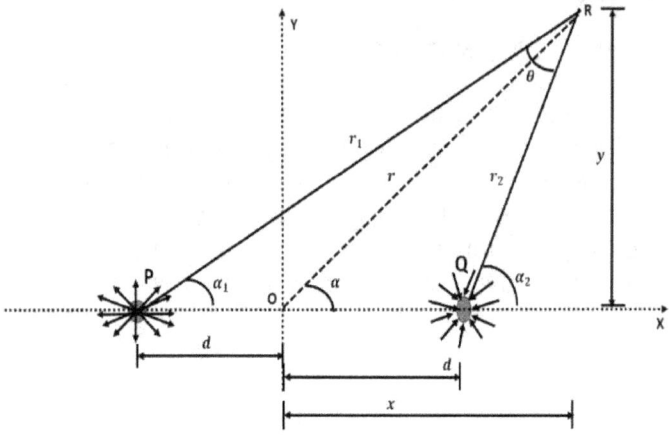

Let R be an arbitrary point in the resultant flow of source and sink of an uniform flow as shown in figure. Pre-assume the following parameters in order to derive the expression for the resultant stream function.

r, α = cylindrical coordinates of point R w.r.t. origin O.

r_1, α_1 = position of the point R w.r.t. source placed at point P.

r_2, α_2 = position of the point R w.r.t. sink placed at point Q.

$k, -k$ = strength of a source and sink placed at the point P and Q.

The stream function due to the source placed at the point P is defined as

INTRODUCTION TO FLUID KINEMATICS

$$\psi_1 = \frac{k}{2\pi}.\alpha_1$$

The stream function due to the sink placed at the point Q is defined as

$$\psi_2 = -\frac{k}{2\pi}.\alpha_2$$

The resultant stream function (ψ) will be the sum of these two stream function and is given by

$$\psi = \psi_1 + \psi_2$$

$$\Rightarrow \psi = \frac{k}{2\pi}.\alpha_1 - \frac{k}{2\pi}.\alpha_2$$

$$\Rightarrow \psi = -\frac{k}{2\pi}(\alpha_2 - \alpha_1) = -\frac{k}{2\pi}.\theta \quad [\because \theta = \alpha_2 - \alpha_1]$$

For a given stream line, ψ = constant. Since the term $\frac{k}{2\pi}$ = constant, then angle $\theta = \alpha_2 - \alpha_1$ will be constant for various positions of R in the reference plane. To satisfy this, the locus of R is a circle with PQ as chord, having its centre on y-axis. This means that the resultant stream lines will be circular arc passing through source and sink of an uniform flow.

Recall the equation for resultant stream function as

$$\psi = -\frac{k}{2\pi}(\alpha_2 - \alpha_1)$$

$$\Rightarrow \psi = \frac{k}{2\pi}(\alpha_1 - \alpha_2)$$

$$\Rightarrow \alpha_1 - \alpha_2 = \frac{2\pi\psi}{k}$$

$$\Rightarrow \tan(\alpha_1 - \alpha_2) = \tan\left(\frac{2\pi\psi}{k}\right)$$

INTRODUCTION TO FLUID KINEMATICS

$$\Rightarrow \frac{\tan\alpha_1 - \tan\alpha_2}{1 + \tan\alpha_1 \tan\alpha_2} = \tan\left(\frac{2\pi\psi}{k}\right) \quad \text{-------(1)}$$

From figure, we have

$$\tan\alpha_1 = \frac{y}{x+d} \quad \text{and} \quad \tan\alpha_2 = \frac{y}{x-d}$$

Put the values of $\tan\alpha_1$ and $\tan\alpha_2$ in the equation (1), we get

$$\frac{\frac{y}{x+d} - \frac{y}{x-d}}{1 + \left(\frac{y}{x+d}\right)\left(\frac{y}{x-d}\right)} = \tan\left(\frac{2\pi\psi}{k}\right)$$

$$\Rightarrow \frac{y(x-d) - y(x+d)}{x^2 - d^2 + y^2} = \tan\left(\frac{2\pi\psi}{k}\right)$$

$$\Rightarrow \frac{-2ay}{x^2 - d^2 + y^2} = \tan\left(\frac{2\pi\psi}{k}\right) \quad \text{------(2)}$$

$$\Rightarrow \psi = -\frac{k}{2\pi}\tan^{-1}\left(\frac{2ay}{x^2 + y^2 - d^2}\right)$$

We translate to cylindrical coordinates, we get

$$\psi = -\frac{k}{2\pi}\tan^{-1}\left(\frac{2.r.d.\sin\alpha}{r^2 - d^2}\right) \quad \text{-----(3)}$$

Let us consider that the distance between source from the origin is d and from the origin to the sink approaches zero as shown in figure. Moreover, we knew that $\tan^{-1}\alpha$ tends to α for very small values of α in radians. So, when distance d close to zero, then equation (3) reduced to

$$\psi : -\frac{d.k.r\sin\alpha}{\pi(r^2 - d^2)} \quad \text{-------(4)}$$

In case of doublet flow, it has observed that both source and sink approaches each other while maintaining the same source and sink strengths. Thus, the product $d.k$ consider as doublet strength remains constant. However, when distance between source and sink vanish, i.e., $d = 0$, then there is no resultant flow. In that case, $r >>> d$ and arbitrary point R approaches the origin as a result equation (4) reduces to

$$\psi = -\frac{d.k}{\pi} \cdot \frac{\sin \alpha}{r}$$

$$\Rightarrow \psi = -D \frac{\sin \alpha}{r} \quad \left[\because D = \frac{d.k}{\pi} = \text{constant} \right]$$

Similarly, we can obtain the expression for velocity potential of doublet along x-axis as

$$\phi = D \frac{\cos \alpha}{r}$$

Let us represent streamlines and velocity potential lines for a doublet which is aligned at some angle from the x-axis.

Let k and $(-k)$ be the strength of source and sink of an uniform flow respectively. Let $2d$ be the distance between them and P be an arbitrary point in the field of doublet flow as shown in figure.

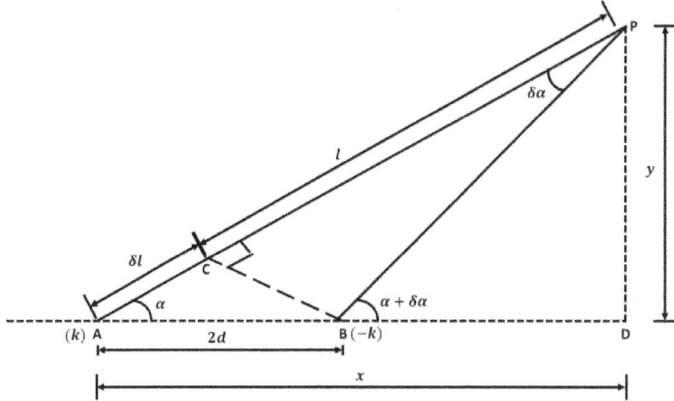

Let α be the angle formed by P at A whereas $\alpha + \delta \alpha$ be the angle formed by P at B. Then doublet strength, denoted by D which is defined as

INTRODUCTION TO FLUID KINEMATICS

$$D = 2d.k \qquad \text{-------(1)}$$

The stream function at P is defined as

$$\psi = \frac{k}{2\pi}\alpha - \frac{k}{2\pi}(\alpha + \delta\alpha)$$

$$\Rightarrow \psi = -\frac{k}{2\pi}\delta\alpha \qquad \text{-------(2)}$$

From B, draw $BC \perp AP$. Then $\angle BPC = \delta\alpha$.

Let $CP = l, AC = \delta l, AP = l + \delta l$. From rt. Angled triangle BPC, $BC = l.\delta\alpha$

From rt. Angled triangle ABC, $BC = 2d.\sin\alpha$

Equating both the values of BC, we get

$$l.\delta\alpha = 2d.\sin\alpha$$

$$\Rightarrow \delta\alpha = \frac{2d.\sin\alpha}{l}$$

Put the value of $\delta\alpha$ in the equation (2), we get

$$\psi = -\frac{k}{2\pi} \cdot \frac{2d\sin\alpha}{l}$$

$$\Rightarrow \psi = -\frac{D}{2\pi} \cdot \frac{\sin\alpha}{l} \qquad [\because D = 2d.k]$$

In doublet flow, $2d \to 0 \Rightarrow \delta\alpha \ll 1$ and $\delta l \to 0 \Rightarrow AP = l$. From the figure, we obtain as

$$\sin\alpha = \frac{PD}{AP} = \frac{y}{l} \text{ and } AP^2 = AD^2 + PD^2 \Rightarrow l^2 = x^2 + y^2$$

Then the expression for stream function, ψ reduced to

INTRODUCTION TO FLUID KINEMATICS

$$\psi = -\frac{D}{2\pi}\cdot\frac{1}{l}\cdot\frac{y}{l} = -\frac{Dy}{2\pi l^2} = -\frac{Dy}{2\pi(x^2+y^2)}$$

$$\Rightarrow x^2 + y^2 = -\frac{Dy}{2\pi\psi}$$

$$\Rightarrow x^2 + y^2 + \frac{Dy}{2\pi\psi} = 0$$

$$\Rightarrow x^2 + y^2 + \frac{Dy}{2\pi\psi} + \left(\frac{D}{4\pi\psi}\right)^2 - \left(\frac{D}{4\pi\psi}\right)^2 = 0$$

$$\Rightarrow x^2 + y^2 + 2\times y \times \frac{D}{4\pi\psi} + \left(\frac{D}{4\pi\psi}\right)^2 - \left(\frac{D}{4\pi\psi}\right)^2 = 0$$

$$\Rightarrow x^2 + \left(y + \frac{D}{4\pi\psi}\right)^2 = \left(\frac{D}{4\pi\psi}\right)^2 \qquad \text{------(3)}$$

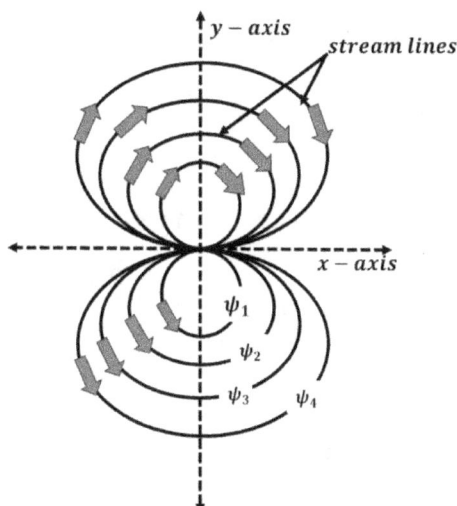

The equation (3) represents a circle with centre $\left(0, \dfrac{D}{4\pi\psi}\right)$ and radius $\dfrac{D}{4\pi\psi}$. the centre of circle lies on the y-axis at a distance of $\dfrac{D}{4\pi\psi}$ from the x-axis. As the radius of the circle is equal to $\dfrac{D}{4\pi\psi}$. Thus, the circle will be tangent to the x-axis. Hence, stream lines of the doublet will be family of circles tangent to the x-axis as shown in figure.

The velocity potential function at P is defined as

$$\phi = \dfrac{k}{2\pi}\log_e(l+\delta l) - \dfrac{k}{2\pi}\log_e l$$

$$\Rightarrow \phi = \dfrac{k}{2\pi}\log_e\left(\dfrac{l+\delta l}{l}\right)$$

$$\Rightarrow \phi = \dfrac{k}{2\pi}\log_e\left(1+\dfrac{\delta l}{l}\right)$$

$$\Rightarrow \phi = \dfrac{k}{2\pi}\left[\dfrac{\delta l}{l} + \dfrac{1}{2}\left(\dfrac{\delta l}{l}\right)^2 + \dfrac{1}{3}\left(\dfrac{\delta l}{l}\right)^3 + \ldots\right]$$

$$\Rightarrow \phi = \dfrac{k}{2\pi}\cdot\dfrac{\delta l}{l} \quad\text{-------(4)}$$

$$\left[\because \dfrac{\delta l}{l} <<< 1,\text{ thus, neglected }\left(\dfrac{\delta l}{l}\right)^2 \text{ and its higher order terms}\right]$$

From the rt. angled triangle ABC, we get

$$\dfrac{\delta l}{2d} = \cos\alpha \Rightarrow \delta l = 2d.\cos\alpha$$

Put the value of δl in the equation (4), we get

$$\phi = \dfrac{k}{2\pi}\cdot\dfrac{2d\cos\alpha}{l}$$

$$\Rightarrow \phi = \dfrac{D}{2\pi}\cdot\dfrac{\cos\alpha}{l} \quad [\because D = 2d.k]$$

In doublet flow, $2d \to 0 \Rightarrow \delta\alpha <<< 1$ and $\delta l \to 0 \Rightarrow AP = l$. From figure, we obtain as

INTRODUCTION TO FLUID KINEMATICS

$$\cos\alpha = \frac{AD}{AP} = \frac{x}{l} \text{ and } AP^2 = AD^2 + PD^2 \Rightarrow l^2 = x^2 + y^2$$

Then the expression for velocity potential, ϕ reduced to

$$\phi = \frac{D}{2\pi} \cdot \frac{1}{l} \cdot \frac{x}{l} = \frac{D}{2\pi} \cdot \frac{x}{l^2}$$

$$\Rightarrow \phi = \frac{Dx}{2\pi(x^2 + y^2)}$$

$$\Rightarrow x^2 + y^2 = \frac{Dx}{2\pi\phi}$$

$$\Rightarrow x^2 + y^2 - \frac{Dx}{2\pi\phi} = 0$$

$$\Rightarrow x^2 - \frac{D}{2\pi} \cdot \frac{x}{\phi} + \left(\frac{D}{4\pi\phi}\right)^2 - \left(\frac{D}{4\pi\phi}\right)^2 + y^2 = 0$$

$$\Rightarrow \left(x - \frac{D}{4\pi\phi}\right)^2 + y^2 = \left(\frac{D}{4\pi\phi}\right)^2 \quad \text{------(5)}$$

The equation (5) is the equation of a circle with centre $\left(\frac{D}{4\pi\phi}, 0\right)$ and radius $\frac{D}{4\pi\phi}$. The centre of the circle lies on x-axis at a distance of $\frac{D}{4\pi\phi}$ from y-axis. As the radius of the circle is equal to the distance of the centre of the circle from the y-axis, thus the circle will be tangent to the y-axis. The potential lines of the doublet will be family of circles tangent to the y-axis as shown in figure.

INTRODUCTION TO FLUID KINEMATICS

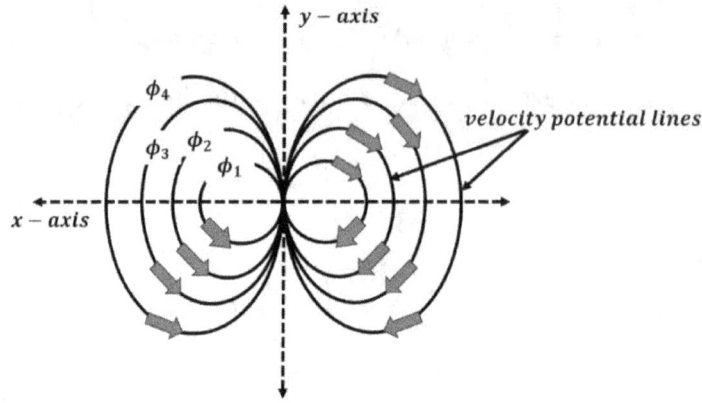

Example: A source and a sink of strength 3 m² / sec and 6 m² / sec are located at $(-1 \ 0)$ and $(1 \ 0)$ and respectively. Determine the velocity at the point $(1 \ 1)$.

Solution:

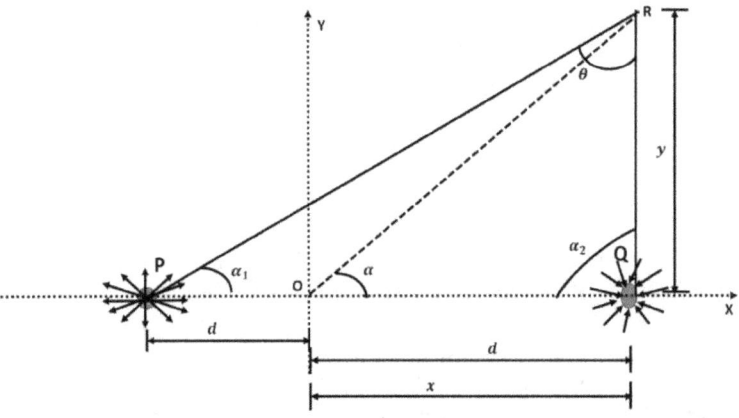

Here, in this case, a source and a sink of strength 3 m² / sec and 6 m² / sec placed at the point $P(-1 \ 0)$ and $Q(1 \ 0)$ respectively. Thus, distance of the source and sink from the origin, $d = 1$ unit. The position of point R is $(x, y) = (1, 1)$.

From figure, it is clear that $\angle \alpha_2 = 90°$ and angle α_1 can be calculated from right angled triangle PQR.

INTRODUCTION TO FLUID KINEMATICS

$$\tan \alpha_1 = \frac{QR}{PQ} = \frac{1}{2} = 0.5$$

$$\alpha_1 = \tan^{-1}(0.5) = 26.56° = 26.56 \times \frac{\pi}{180} \text{ radians}$$

$$\Rightarrow \alpha_1 = 0.463 \text{ radians}$$

Also, angle $\alpha_2 = 90° = 90 \times \frac{\pi}{180}$ radians $= \frac{\pi}{2}$ radians

To determine the velocity at the point R, let us first calculate the stream function in terms of x and y coordinates. The stream function defined as

$$\psi = \frac{k_1}{2\pi} \alpha_1 - \frac{k_2}{2\pi} \alpha_2$$

$$\Rightarrow \psi = \frac{k_1}{2\pi} \tan^{-1}\left(\frac{y}{x+d}\right) - \frac{k_2}{2\pi} \tan^{-1}\left(\frac{y}{x-d}\right)$$

The velocity component $u = \frac{\partial \psi}{\partial y}$ and $v = -\frac{\partial \psi}{\partial x}$

$$u = \frac{\partial \psi}{\partial y}$$

$$\Rightarrow u = \frac{\partial}{\partial y}\left[\frac{k_1}{2\pi} \tan^{-1}\left(\frac{y}{x+d}\right) - \frac{k_2}{2\pi} \tan^{-1}\left(\frac{y}{x-d}\right)\right]$$

$$\Rightarrow u = \frac{k_1}{2\pi} \cdot \frac{x+d}{(x+d)^2 + y^2} - \frac{k_2}{2\pi} \cdot \frac{x-d}{(x-d)^2 + y^2}$$

Put $x = 1, y = 1$ and $d = 1$ in the above equation. We get

$$u = \frac{k_1}{2\pi} \cdot \frac{1+1}{(1+1)^2 + 1^2} - \frac{k_2}{2\pi} \cdot \frac{1-1}{(1-1)^2 + 1^2}$$

$$\Rightarrow u = \frac{k_1}{2\pi} \cdot \frac{2}{5} - \frac{k_2}{2\pi} \cdot 0$$

$$\Rightarrow u = \frac{k_1}{2\pi} \cdot \frac{2}{5} = \frac{3}{2\pi} \cdot \frac{2}{5} = 0.191 \text{ m/sec}$$

INTRODUCTION TO FLUID KINEMATICS

$$v = -\frac{\partial \psi}{\partial x}$$

$$\Rightarrow v = -\frac{\partial}{\partial x}\left[\frac{k_1}{2\pi}\tan^{-1}\left(\frac{y}{x+d}\right) - \frac{k_2}{2\pi}\tan^{-1}\left(\frac{y}{x-d}\right)\right]$$

$$\Rightarrow v = -\frac{k_1}{2\pi}\cdot\frac{y}{(x+d)^2+y^2} - \frac{k_2}{2\pi}\cdot\frac{y}{(x-d)^2+y^2}$$

Put $x=1, y=1$ and $d=1$ in the above equation. We get

$$v = \frac{k_1}{2\pi}\cdot\frac{1}{(1+1)^2+1^2} - \frac{k_2}{2\pi}\cdot\frac{1}{(1-1)^2+1^2}$$

$$\Rightarrow v = \frac{k_1}{2\pi}\cdot\frac{1}{5} - \frac{k_2}{2\pi}\cdot\frac{1}{1}$$

$$\Rightarrow v = \frac{k_1}{2\pi}\cdot\frac{1}{5} - \frac{k_2}{2\pi}$$

Put $k_1 = 3, k_2 = 6$ in the above equation. We get

$$v = \frac{3}{2\pi}\cdot\frac{1}{5} - \frac{6}{2\pi} = 0.0956 - 0.9554 = 0.8598 \text{ m/sec}$$

∴ The resultant velocity,

$$V = \sqrt{u^2 + v^2} = \sqrt{(0.191)^2 + (0.8598)^2}$$

$$\Rightarrow V = \sqrt{0.036481 + 0.739256} = \sqrt{0.775737} = 0.881 \text{ m/sec}$$

3. STRESS

INTRODUCTION

Consider the water (fluid) is filled tightly placed between two plates. As the upper plate is pulled with a force F while the lower plate is held fixed, the fluid deforms as shown in Figure.

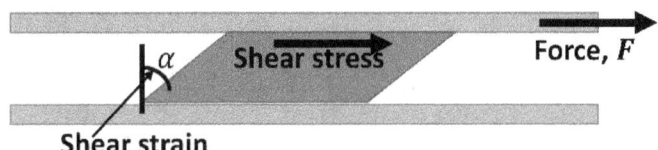

Figure: *Deformation of a fluid filled in between two parallel plates under the action of Force (F)*

Assuming there is no slip between the fluid and the plates. On applying the force F, the upper surface of the fluid is displaced by an amount equal to the displacement of the upper plate while the lower surface remains stationary. The angle of deformation α (known as angular displacement) increases in proportion to the applied force F. In equilibrium state, the net force acting on the upper plate in the horizontal direction must be zero, and thus a force equal and opposite to F must be acted on the plate. The opposing force which develops at the plate-fluid interface due to friction is expressed as $F = \tau A$, where τ is the shear stress and A is the contact area between the upper plate and the fluid. The

INTRODUCTION TO FLUID KINEMATICS

water (fluid) is filled within two large parallel plates, where fluid layer in contact with the upper plate move continuously at the velocity of the plate irrespective of the amount of force F. The fluid velocity would decrease with depth because of friction between fluid layers, reaching zero at the lower plate. Here, **stress** is induced as a result of applied force. Stress can be defined as force per unit area upon which it acts. The normal component of a force acting on a surface per unit area is called **normal stress**, and the tangential component of a force acting on a surface per unit area is called **shear stress** which is defined as

$$\tau = \frac{F}{A}$$

Shear stress is the frictional force per unit area. For an infinitesimal fluid element, we have

$$\tau = \frac{v}{y}$$

$$\Rightarrow \tau = \mu \left(\frac{\partial v}{\partial y}\right)$$

which is known as **Newton's Law of viscosity.** μ is a constant of proportionality, which is called the *co-efficient of dynamic viscosity.* This parameter depends on the pressure and temperature but does not depends on the pressure in case of gases. The ratio of μ to the density ρ of the fluid is known as *co-efficient of kinematic viscosity* and defined as

$$v = \frac{\mu}{\rho}$$

The SI unit of **stress** is Newton per square metre (N/m²) or Pascal (Pa).

Let us discuss the various types of stress induced in solids and liquids as well.

CAUCHY'S STRESS POSTULATES

1. The stress vector remains unchanged for all surfaces passing through any fixed point and having same normal vector $\hat{n}$ at the fixed point. It states that the stress vector is depends on the normal vector $\hat{n}$ and is unaffected by the shape of the fluid element. Mathematically, we can express as $\vec{T} = \vec{T}(\hat{n})$

2. The stress vector acting on any particular point of the same surface are equal in magnitude and opposite to each other. $\vec{T}(K,\hat{m}) = \vec{T}(K,-\hat{m})$

TYPES OF STRESS

NORMAL STRESS

The force per unit area which acts perpendicular to the surface of the body is known as **Normal stress.** It has two category namely **Tensile stress** and **Compressive stress.** Under the action of Normal stress, the body deformed such that the net force acting on the body is zero. Moreover, Tensile stress and Compressive stress consider as **Longitudinal stress** for engineering purposes.

TENSILE STRESS
The equal and opposite forces applied on a solid object to increase its length, a restoring force created which is equal to the applied force F acts perpendicular to the cross-sectional area of the object is known as **Tensile stress.** Therefore, *Tensile stress is defined as restoring force or deforming force per unit area which acts perpendicular to the cross-section of the body.*

INTRODUCTION TO FLUID KINEMATICS

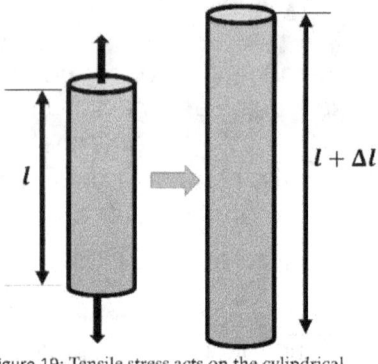

Figure 19: Tensile stress acts on the cylindrical rod results into the elongation of length in one-dimensional.

Figure shows the increment in length of a solid object when two equal and opposite forces are applied at the ends of a rod. Here, induced force equivalent to the applied force which acts perpendicular to the cross-sectional area of object is called as **Tensile Stress.** It is defined as

$$Tensile\ stress = \frac{Applied\ Force}{Area} = \frac{F}{A}$$

COMPRESSIVE STRESS

The equal and opposite forces applied on a solid object to decrease its length, a restoring force created which is equal to the applied force F acts perpendicular to the cross-sectional area of the object is known as **Compressive stress.** Therefore, *Compressive stress is defined as restoring force or deforming force per unit area which acts perpendicular to the cross-section of the body.*

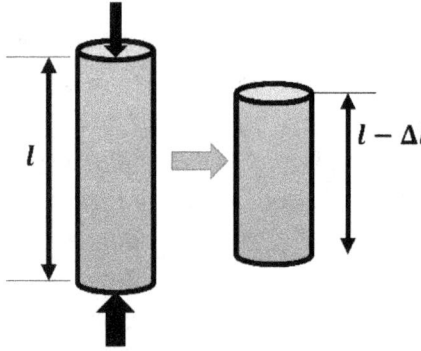

Figure 20: Compressive stress acts on the cylindrical rod results into the reduction of length in one-dimensional.

Figure shows the reduction in length of a solid object when two equal and opposite forces are applied at the ends of a rod. Here, induced force equivalent to the applied force which acts perpendicular to the cross-sectional area of object is called as **Compressive Stress.** It is defined as

$$Compressive\ stress = \frac{Applied\ Force}{Area} = \frac{F}{A}$$

TANGENTIAL STRESS or SHEAR STRESS

The body deform tangentially when two equal and opposite forces act along the tangents to the surfaces of the body. The body deform its shape with reference to static surface of a body. As shown in figure, the body is under the stress known as **Tangential stress** or **Shearing stress.**

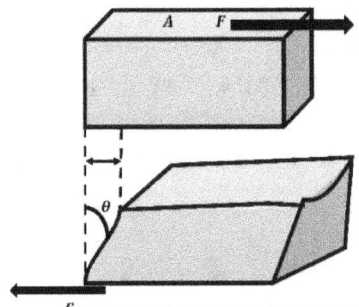

Figure 21: Shear stress act along the tangent to the surface of cuboidal object results in displacement of object from the initial point.

Therefore, **Tangential stress** *is defined as the ratio of the force acts tangent to the surface area.* It is defined as

$$Shearing\ stress = \frac{Force}{surface\ Area} = \frac{F}{A}$$

INTRODUCTION TO FLUID KINEMATICS

BULK STRESS or VOLUME STRESS or HYDRAULIC STRESS

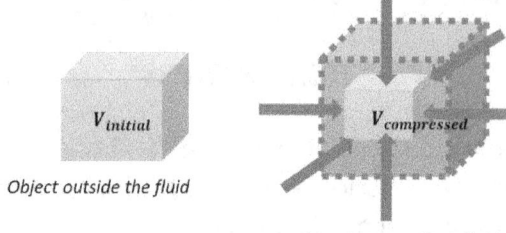

Object outside the fluid

Object immersed in a fluid

The volume of the object immersed in a fluid decrease when the fluid exerts force on the surface of the object as shown in figure. The immersed object is under a stress known as **Bulk stress** or **Volume stress** or **Hydraulic stress.**
Therefore, **Bulk stress** *is defined as the force per unit area acts perpendicular to the surface of the object.* It is defined as

$$Bulk\ stress = \frac{Force}{surface\ Area} = \frac{F}{A}$$

EXPRESSION FOR STRESS TENSOR

Consider a fluid element in tetrahedron shape in three dimensional coordinates planes. The fluid element is considered to be infinitesimal small whose area is A oriented in any direction by a normal unit vector $\hat{n}$ to formulated the expression for stress tensor. The stress vector on this plane denoted by $\vec{F}$. The stress vectors on the tetrahedron's faces are $\vec{F_1}, \vec{F_2},$ and $\vec{F_3}$. The stress tensor denoted by σ whose components considered to be σ_{ij}.

By Newton's 2nd law of motion, we have

$$\vec{F}A - \vec{F_1}A_1 - \vec{F_2}A_2 - \vec{F_3}A_3 = \rho\left(\frac{h}{3}A\right)a \text{----(1)}$$

INTRODUCTION TO FLUID KINEMATICS

where ρ = density of the fluid

a = acceleraton of the fluid flow

h = height of the tetrahedron fluid element from the reference line

Following the areas of the faces that are perpendicular to the axes as:

$$A_1 = (\hat{n}e_1)A = n_1 A$$
$$A_2 = (\hat{n}e_2)A = n_2 A$$
$$A_3 = (\hat{n}e_3)A = n_3 A$$

Now the equation (1) reduces to

$$\vec{F} - \vec{F}_1 n_1 A - \vec{F}_2 n_2 A - \vec{F}_3 n_3 A = \rho\left(\frac{h}{3}A\right)a$$

$$\Rightarrow \vec{F} - \vec{F}_1 n_1 - \vec{F}_2 n_2 - \vec{F}_3 n_3 = \rho\left(\frac{h}{3}\right)a \text{----(2)}$$

Consider the fluid element whose projections are perpendicular to the coordinate axes of a cartesian coordinate system. The stress vectors with respect to faces OBC, OCA, OAB and ABC are $\vec{F}_x, \vec{F}_y, \vec{F}_z$ and $\vec{F}_t$ respectively. Then equation (2) modify as

$$\vec{F}_t - \vec{F}_x l - \vec{F}_y m - \vec{F}_z n = \rho\left(\frac{h}{3}\right)a \text{----(3)}$$

Assume that the face ABC approach 0 so,

$$\rho\left(\frac{h}{3}\right)a \to 0$$

$\Rightarrow$ R.H.S of Eq (3) = 0

$\Rightarrow \vec{F}_t - \vec{F}_x l - \vec{F}_y m - \vec{F}_z n = 0$

$\Rightarrow \vec{F}_t = \vec{F}_x l + \vec{F}_y m + \vec{F}_z n$ ----(4)

INTRODUCTION TO FLUID KINEMATICS

Let $\sigma_{nx}, \sigma_{ny}, \sigma_{nz}$ be the cartesian components of $\vec{F}_t$ and $\hat{i}, \hat{j}, \hat{k}$ are the unit vectors parallel to the axes as shown in figure.

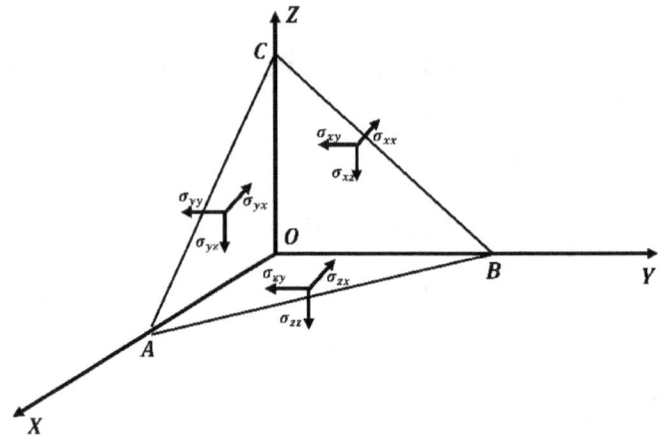

Then we have

$$\vec{F}_t = \sigma_{nx}\hat{i} + \sigma_{ny}\hat{j} + \sigma_{nz}\hat{k} \quad ----(5)$$

The components of stress vectors are defined as

$$\left.\begin{array}{l} \vec{F}_x = \sigma_{xx}\hat{i} + \sigma_{xy}\hat{j} + \sigma_{xz}\hat{z} \\ \vec{F}_y = \sigma_{yx}\hat{i} + \sigma_{yy}\hat{j} + \sigma_{yz}\hat{z} \\ \vec{F}_z = \sigma_{zx}\hat{i} + \sigma_{zy}\hat{j} + \sigma_{zz}\hat{z} \end{array}\right\} ---(6)$$

Using equation (4), (5) and (6), we obtain the co-efficient of $\hat{i}, \hat{j}, \hat{k}$ as

$$\left.\begin{array}{l} \sigma_{nx} = \sigma_{xx}l + \sigma_{yx}m + \sigma_{zx}n \\ \sigma_{ny} = \sigma_{xy}l + \sigma_{yy}m + \sigma_{zy}n \\ \sigma_{nz} = \sigma_{xz}l + \sigma_{yz}m + \sigma_{zz}n \end{array}\right\} ---(7)$$

INTRODUCTION TO FLUID KINEMATICS

The above relation can be express in matrix form as

$$\begin{bmatrix} \sigma_{nx} \\ \sigma_{ny} \\ \sigma_{nz} \end{bmatrix} = \begin{bmatrix} \sigma_{xx} & \sigma_{yx} & \sigma_{zx} \\ \sigma_{xy} & \sigma_{yy} & \sigma_{zy} \\ \sigma_{xz} & \sigma_{yz} & \sigma_{zz} \end{bmatrix} \begin{bmatrix} l \\ m \\ n \end{bmatrix}$$

SYMMETRY OF STRESS TENSOR

Let us consider an elementary rectangular parallelepiped with $K(x,y,z)$ as centre and edges of lengths $\Delta x, \Delta y, \Delta z$ w.r.t. standard orthogonal coordinate axes as shown in figure. Assume that the fluid element is moving with the fluid and its mass $\rho \Delta x \Delta y \Delta z$ remains constant. Since fluid element with the position $K(x,y,z)$ moving with the uniform fluid motion, so the coordinates of K_1 and K_2 are $\left(x-\dfrac{\Delta x}{2}, y, z\right)$ and $\left(x+\dfrac{\Delta x}{2}, y, z\right)$ respectively.

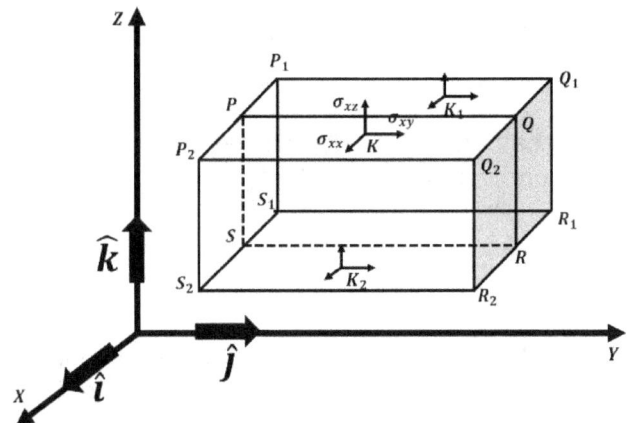

At K, the force components parallel to OX, OY, OZ on the rectangular surface $PQRS$ of area $\Delta y \Delta z$ through K are
$\left[\sigma_{xx}\Delta y\Delta z, \ \sigma_{xy}\Delta y\Delta z, \ \sigma_{xz}\Delta y\Delta z\right]$

At K_1, the force components on the rectangular surface

INTRODUCTION TO FLUID KINEMATICS

$P_1Q_1R_1S_1$ (parallel to $PQRS$) of area $\Delta y \Delta z$ are

$$\left[-\left(\sigma_{xx}-\frac{\Delta x}{2}\cdot\frac{\delta\sigma_{xx}}{\delta x}\right)\Delta y\Delta z,-\left(\sigma_{xy}-\frac{\Delta x}{2}\cdot\frac{\delta\sigma_{xy}}{\delta x}\right)\Delta y\Delta z,-\left(\sigma_{xz}-\frac{\Delta x}{2}\cdot\frac{\delta\sigma_{xz}}{\delta x}\right)\Delta y\Delta z\right]$$

Here $-ve$ sign indicates that the unit normal vector acts outward as fluid element is moving with uniform fluid motion.

At K_2, the force components on the rectangular surface $P_2Q_2R_2S_2$ (parallel to $PQRS$) of area $\Delta y \Delta z$ are

$$\left[\left(\sigma_{xx}+\frac{\Delta x}{2}\cdot\frac{\delta\sigma_{xx}}{\delta x}\right)\Delta y\Delta z,\left(\sigma_{xy}+\frac{\Delta x}{2}\cdot\frac{\delta\sigma_{xy}}{\delta x}\right)\Delta y\Delta z,\left(\sigma_{xz}+\frac{\Delta x}{2}\cdot\frac{\delta\sigma_{xz}}{\delta x}\right)\Delta y\Delta z\right]$$

Thus, the forces on the parallel planes $P_1Q_1R_1S_1$ and $P_2Q_2R_2S_2$ passing through K_1 and K_2 are equivalent to a single force at K with components

$$\left[\frac{\delta\sigma_{xx}}{\delta x}\Delta x\Delta y\Delta z,\frac{\delta\sigma_{xy}}{\delta x}\Delta x\Delta y\Delta z,\frac{\delta\sigma_{xz}}{\delta x}\Delta x\Delta y\Delta z\right] \text{ together with}$$

couples whose moments are $-\sigma_{xz}\Delta x\Delta y\Delta z$ about OY and $\sigma_{xy}\Delta x\Delta y\Delta z$ about OZ.

Similarly, the forces on the parallel planes perpendicular to $y-$ axis are equivalent to a single force at K with components

$$\left[\frac{\delta\sigma_{yx}}{\delta y}\Delta x\Delta y\Delta z,\frac{\delta\sigma_{yy}}{\delta y}\Delta x\Delta y\Delta z,\frac{\delta\sigma_{yz}}{\delta y}\Delta x\Delta y\Delta z\right] \text{ together}$$

with couples whose moments are $-\sigma_{yx}\Delta x\Delta y\Delta z$ about OZ and $\sigma_{yz}\Delta x\Delta y\Delta z$ about OX.

Again, the forces on the parallel planes perpendicular to $z-$ axis are equivalent to a single force at K with components

$$\left[\frac{\delta\sigma_{zx}}{\delta z}\Delta x\Delta y\Delta z,\frac{\delta\sigma_{zy}}{\delta z}\Delta x\Delta y\Delta z,\frac{\delta\sigma_{zz}}{\delta z}\Delta x\Delta y\Delta z\right] \text{ together with}$$

couples whose moments are $-\sigma_{zy}\Delta x\Delta y\Delta z$ about OX and

INTRODUCTION TO FLUID KINEMATICS

$\sigma_{zx}\Delta x\Delta y\Delta z$ about OY.

Thus, the surfaces forces on all six faces of the rectangular parallelopiped are equivalent to a single force at K having components

$$\left[\left(\frac{\delta\sigma_{xx}}{\delta x}+\frac{\delta\sigma_{yx}}{\delta y}+\frac{\delta\sigma_{zx}}{\delta z}\right)\Delta x\Delta y\Delta z,\right.$$

$$\left(\frac{\delta\sigma_{xy}}{\delta x}+\frac{\delta\sigma_{yy}}{\delta y}+\frac{\delta\sigma_{zy}}{\delta z}\right)\Delta x\Delta y\Delta z,$$

$$\left.\left(\frac{\delta\sigma_{xz}}{\delta x}+\frac{\delta\sigma_{yz}}{\delta y}+\frac{\delta\sigma_{zz}}{\delta z}\right)\Delta x\Delta y\Delta z\right]$$

together with a moments couple having components

$$\left[\left(\sigma_{yz}-\sigma_{zy}\right)\Delta x\Delta y\Delta z,\left(\sigma_{zx}-\sigma_{xz}\right)\Delta x\Delta y\Delta z,\left(\sigma_{xy}-\sigma_{yx}\right)\Delta x\Delta y\Delta z\right].$$

Consider the moments about the $\hat{i}$ direction through K, we get

Total moments = (moment of inertia about OX) × (angular acceleration)

$$\Rightarrow \left(\sigma_{yz}-\sigma_{zy}\right)\Delta x\Delta y\Delta z + e_4 = e_5$$

Where e_4 and e_5 are the fourth and fifth order of extension of $\Delta x, \Delta y, \Delta z$. We only consider extension up to third order of $\Delta x, \Delta y, \Delta z$, we obtain as

$$\left(\sigma_{yz}-\sigma_{zy}\right)\Delta x\Delta y\Delta z = 0$$

Since $\Delta x, \Delta y, \Delta z$ are arbitrary and infinitesimal small, so coefficient of $\Delta x\Delta y\Delta z$ must vanish and we obtain as $\sigma_{yz}-\sigma_{zy}=0$. Hence, we get

$$\sigma_{yz}=\sigma_{zy}$$

Similarly, we can prove that $\sigma_{zx} = \sigma_{xz}$ and $\sigma_{xy} = \sigma_{yx}$

Thus, above expressions shows that the stress tensor is symmetric.

PRINCIPAL STRESSES

When fluid is in flowing state, then pressure acts inward and normal to the surface, along with this shear stress also acts on fluid flow due to the viscosity. Thus, the stress tensor can be express as

$$\sigma_{ij} = \begin{pmatrix} \sigma_{xx} & \sigma_{xy} & \sigma_{xz} \\ \sigma_{yx} & \sigma_{yy} & \sigma_{yz} \\ \sigma_{zx} & \sigma_{zy} & \sigma_{zz} \end{pmatrix} = \begin{pmatrix} -P & 0 & 0 \\ 0 & -P & 0 \\ 0 & 0 & -P \end{pmatrix} + \begin{pmatrix} \tau_{xx} & \tau_{xy} & \tau_{xz} \\ \tau_{yx} & \tau_{yy} & \tau_{yz} \\ \tau_{zx} & \tau_{zy} & \tau_{zz} \end{pmatrix}$$

Where principal stress, P are the normal stress acts on fluid surface with zero shear stress.

However, in case of static fluid state, there are only normal stress acts on the fluid's surface, which is considered as pressure, P in inward direction. Thus, the stress tensor w.r.t. three dimensional coordinate system reduces to

$$\sigma_{ij} = \begin{pmatrix} \sigma_{xx} & \sigma_{xy} & \sigma_{xz} \\ \sigma_{yx} & \sigma_{yy} & \sigma_{yz} \\ \sigma_{zx} & \sigma_{zy} & \sigma_{zz} \end{pmatrix} = \begin{pmatrix} -P & 0 & 0 \\ 0 & -P & 0 \\ 0 & 0 & -P \end{pmatrix}$$

We can define pressure for an incompressible fluid as

$$P = -\frac{1}{3}(\sigma_{xx} + \sigma_{yy} + \sigma_{zz})$$

DIAGONALIZATION OF STRESS TENSOR

As we know that the nine components of stress tensor, σ_{ij} at any point are necessary to calculate to know the state of stress at a point in any particular flow field. The stress tensor defined as

$$\begin{bmatrix} \sigma_1 \\ \sigma_2 \\ \sigma_3 \end{bmatrix} = \begin{bmatrix} \sigma_{xx} & \sigma_{yx} & \sigma_{zx} \\ \sigma_{xy} & \sigma_{yy} & \sigma_{zy} \\ \sigma_{xz} & \sigma_{yz} & \sigma_{zz} \end{bmatrix} \begin{bmatrix} l \\ m \\ n \end{bmatrix}$$

Let us assume that $\sigma_{xz} = \sigma_{yz} = \sigma_{zz} = 0$

Since stress is a symmetric tensor of order two. Thus, we have $\sigma_{yz} = \sigma_{zy} = 0, \sigma_{zx} = \sigma_{xz} = 0$

Here, state of stress is given by

$$\left. \begin{array}{l} \sigma_1 = l\sigma_{xx} + m\sigma_{yx} \\ \sigma_2 = l\sigma_{xy} + m\sigma_{yy} \\ \sigma_3 = 0 \end{array} \right\} \quad \text{------(1)}$$

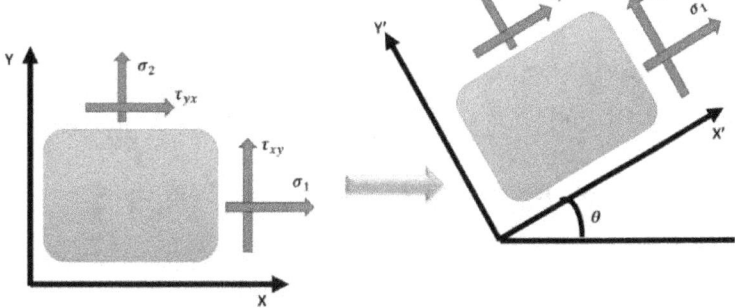

Consider a fluid element positioned at point P with normal $\hat{n}$ to the XY plane and normal $\hat{n}$ makes an angle θ with $x-$axis as shown in figure. Then $l = \cos\theta$ and $m = \sin\theta$, equ (1) becomes

$$\sigma_1 = \sigma_{xx}\cos\theta + \sigma_{yx}\sin\theta$$
$$\sigma_2 = \sigma_{xy}\cos\theta + \sigma_{yy}\sin\theta$$
$$\sigma_3 = 0$$

INTRODUCTION TO FLUID KINEMATICS

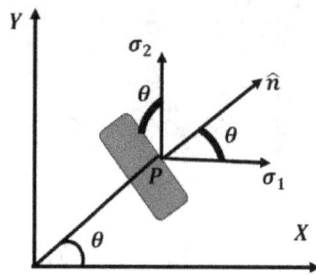

The normal and tangential components of stress are given by

$$\sigma_n = \sigma_1 \cos\theta + \sigma_2 \sin\theta$$
$$= (\sigma_{xx}\cos\theta + \sigma_{yx}\sin\theta)\cos\theta + (\sigma_{xy}\cos\theta + \sigma_{yy}\sin\theta)\sin\theta$$
$$= \sigma_{xx}\cos^2\theta + \sigma_{yy}\sin^2\theta + 2\sigma_{xy}\sin\theta\cos\theta$$
$$= \left(\frac{\sigma_{xx}+\sigma_{yy}}{2}\right) + \left(\frac{\sigma_{xx}-\sigma_{yy}}{2}\right)\cos 2\theta + \sigma_{xy}\sin 2\theta$$

$$\tau_i = \sigma_2 \cos\theta - \sigma_1 \sin\theta$$
$$= (\sigma_{xy}\cos\theta + \sigma_{yy}\sin\theta)\cos\theta - (\sigma_{xx}\cos\theta + \sigma_{yx}\sin\theta)\sin\theta$$
$$= (\sigma_{yy} - \sigma_{xx})\sin\theta\cos\theta + \sigma_{xy}\cos^2\theta - \sigma_{yx}\sin^2\theta$$
$$= -\left(\frac{\sigma_{xx}-\sigma_{yy}}{2}\right)\sin 2\theta + \sigma_{xy}\cos 2\theta$$

Principal stress refer to zero shear stress in the principal plane. Thus, there are biaxial forces acts on two planes on which the shear stress value is zero. In order to determine the position of this principal plane relative to $XY-$ plane, put $\tau_i = 0$ in the equation of tangential component of stress and we obtain as

INTRODUCTION TO FLUID KINEMATICS

$$\tau_i = 0 = -\left(\frac{\sigma_{xx} - \sigma_{yy}}{2}\right)\sin 2\theta + \sigma_{xy}\cos 2\theta$$

$$\Rightarrow \left(\frac{\sigma_{xx} - \sigma_{yy}}{2}\right)\sin 2\theta = \sigma_{xy}\cos 2\theta$$

$$\Rightarrow \frac{\sin 2\theta}{\cos 2\theta} = \frac{2\sigma_{xy}}{\sigma_{xx} - \sigma_{yy}}$$

$$\Rightarrow \tan 2\theta = \frac{2\sigma_{xy}}{\sigma_{xx} - \sigma_{yy}}$$

Since $\tan 2\theta = \tan 2\left(\frac{\pi}{2} + \theta\right)$ is true, then $\tan 2\theta = \dfrac{2\sigma_{xy}}{\sigma_{xx} - \sigma_{yy}}$

holds for two values of θ.

Hence, there are two mutually perpendicular directions of principal planes of stress at any point where tangential components are vanishes.

The two directions calculated by two values of θ are refer to principal planes of stress at particular point and normal stress corresponding to them are known as the principal stresses.

TRANSFORMATION OF STRESS COMPONENTS

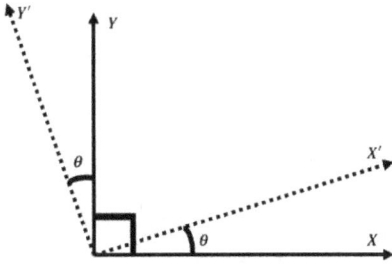

Let the two-dimensional stress components σ_{xx}, σ_{yy} and σ_{xy} at O w.r.t. coordinate axes OX and OY. Let OX', OY' be another set of orthogonal axes as shown in figure. Then $\sigma_{x'x'}, \sigma_{y'y'}$ and $\sigma_{x'y'}$ be the stress components w.r.t. new coordinate axes OX' and OY'. The direction cosines of axes w.r.t. new coordinate axes are given as

	OX	OY

INTRODUCTION TO FLUID KINEMATICS

OX'	$l_1 = \cos\theta$	$m_1 = \cos\left(\dfrac{\pi}{2} - \theta\right) = \sin\theta$
OY'	$l_2 = \cos\left(\dfrac{\pi}{2} + \theta\right)$ $\Rightarrow l_2 = -\sin\theta$	$m_2 = \cos\theta$

The stress components in x, y directions on the elementary area normal to OX', OY' are given by

$$\sigma_{x'x} = l_1\sigma_{xx} + m_1\sigma_{yx}$$
$$\sigma_{x'y} = l_1\sigma_{xy} + m_1\sigma_{yy}$$
$$\sigma_{y'x} = l_2\sigma_{xx} + m_2\sigma_{yx}$$
$$\sigma_{y'y} = l_2\sigma_{xy} + m_2\sigma_{yy}$$

The transformed stress components in x', y' directions on the elementary area normal to OX', OY' are given by

$$\sigma_{x'x'} = l_1\sigma_{x'x} + m_1\sigma_{x'y}$$
$$\sigma_{x'y'} = l_2\sigma_{x'x} + m_2\sigma_{x'y}$$
$$\sigma_{y'y'} = l_2\sigma_{y'x} + m_2\sigma_{y'y}$$

Using above relations, new stress components in x', y' direction reduced to

$$\sigma_{x'x'} = l_1^2\sigma_{xx} + m_1^2\sigma_{yy} + 2l_1\sigma_{xy}$$
$$\sigma_{x'y'} = l_1l_2\sigma_{xx} + m_1m_2\sigma_{yy} + (l_1m_2 + l_2m_1)\sigma_{xy}$$
$$\sigma_{y'y'} = l_2^2\sigma_{xx} + m_2^2\sigma_{yy} + 2l_2m_2\sigma_{xy}$$

Using the above mentioned table, above expressions can be re-written in terms of angle θ are follows:

INTRODUCTION TO FLUID KINEMATICS

$$\sigma_{x'x'} = \cos^2\theta\sigma_{xx} + \sin^2\theta\sigma_{yy} + 2\sin\theta\cos\theta\sigma_{xy}$$

$$\sigma_{y'y'} = \sin^2\theta\sigma_{xx} + \cos^2\theta\sigma_{yy} - 2\sin\theta\cos\theta\sigma_{xy}$$

$$\sigma_{x'y'} = -\sin\theta\cos\theta\sigma_{xx} + \sin\theta\cos\theta\sigma_{yy} + \left(\cos^2\theta - \sin^2\theta\right)\sigma_{xy}$$

Recall high school knowledge, we have

$$\sin 2\theta = 2\sin\theta\cos\theta$$

$$\sin^2\theta = \frac{1}{2}(1-\cos 2\theta)$$

$$\cos^2\theta = \frac{1}{2}(1+\cos 2\theta)$$

Using above relations, transformed stress components can be re-written as

$$\sigma_{x'x'} = \frac{1}{2}(\sigma_{xx}+\sigma_{yy}) + \frac{1}{2}(\sigma_{xx}-\sigma_{yy})\cos 2\theta + \sigma_{xy}\sin 2\theta \quad \text{------(1)}$$

$$\sigma_{y'y'} = \frac{1}{2}(\sigma_{xx}+\sigma_{yy}) - \frac{1}{2}(\sigma_{xx}-\sigma_{yy})\cos 2\theta - \sigma_{xy}\sin 2\theta \quad \text{------(2)}$$

$$\sigma_{x'y'} = -\frac{1}{2}(\sigma_{xx}-\sigma_{yy})\sin 2\theta + \sigma_{xy}\cos 2\theta \quad \text{--------------(3)}$$

It follows that

$$\sigma_{x'x'} + \sigma_{y'y'} = \sigma_{xx} + \sigma_{yy}$$

$$\sigma_{x'x'}\sigma_{y'y'} - \sigma_{x'y'}^2 = \sigma_{xx}\sigma_{yy} - \sigma_{xy}^2$$

The above expressions shows that stress components are remain invariant for the transformation of coordinate axes, which consider as **stress invariants in two-dimensional coordinate system.**

MOHR'S CIRCLE

Mohr's circle is a graphical representation of the state of stress induced at particular point in respective planes. Mohr's circle indicates the relation between components of stress in two-dimensional and three dimensional coordinate system respectively. It presents the transformation of principal stress and tangentially stress w.r.t coordinate system.

Mohr's circle for a state of stress in 2D coordinate system

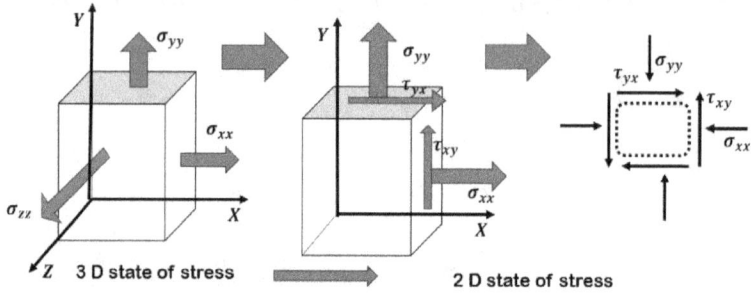

3 D state of stress   2 D state of stress

Let two dimensional stress components σ_{xx}, σ_{yy}, and τ_{xy} at O w.r.t. coordinate axes OX and OY. The fluid element is considered to be infinitesimal small whose normal unit vector $\hat{n}$ makes an angle θ with x-axis. Thus, stress vector express as

$$\sigma_{ij} = \begin{pmatrix} \sigma_{xx} & \tau_{xy} \\ \tau_{xy} & \sigma_{yy} \end{pmatrix} \begin{pmatrix} \cos\theta \\ \sin\theta \end{pmatrix} = \begin{pmatrix} \sigma_{xx}\cos\theta + \tau_{xy}\sin\theta \\ \tau_{xy}\cos\theta + \sigma_{yy}\sin\theta \end{pmatrix}$$

where normal stress is given by

$$\sigma_n = \sigma_{ij}.\hat{n} = \frac{1}{2}(\sigma_{xx} + \sigma_{yy}) + \frac{1}{2}(\sigma_{xx} - \sigma_{yy})\cos 2\theta + \tau_{xy}\sin 2\theta$$

and tangential component of stress is given by

$$\sigma_t = \sigma_{ij}.\hat{m} = \frac{1}{2}(\sigma_{xx} - \sigma_{yy})\sin 2\theta - \tau_{xy}\cos 2\theta \qquad \text{provided}$$

$$\hat{n} = (\cos\theta \quad \sin\theta)^T, \quad \hat{m} = (\sin\theta \quad -\cos\theta)^T$$

Since there are two mutually perpendicular directions of principal planes of stress at any point where tangential components are vanishes. Thus, we have

INTRODUCTION TO FLUID KINEMATICS

$$\sigma_t = 0 \Rightarrow \frac{1}{2}(\sigma_{xx} - \sigma_{yy})\sin 2\theta - \tau_{xy}\cos 2\theta = 0$$

$$\Rightarrow \left(\frac{\sigma_{xx} - \sigma_{yy}}{2}\right)\sin 2\theta = \tau_{xy}\cos 2\theta$$

$$\Rightarrow \tan 2\theta = \frac{\tau_{xy}}{\left(\frac{\sigma_{xx} - \sigma_{yy}}{2}\right)}$$

$$\Rightarrow \sin 2\theta = \pm \frac{1}{\sqrt{1 + \frac{1}{\tan^2 2\theta}}} = \pm \frac{\tau_{xy}}{\sqrt{\left(\frac{\sigma_{xx} - \sigma_{yy}}{2}\right)^2 + \tau_{xy}^2}}$$

$$\cos 2\theta = \pm \frac{1}{\sqrt{1 + \tan^2 2\theta}} = \pm \frac{\frac{1}{2}(\sigma_{xx} - \sigma_{yy})}{\sqrt{\left(\frac{\sigma_{xx} - \sigma_{yy}}{2}\right)^2 + \tau_{xy}^2}}$$

On solving the trigonometric equations, we get two solutions gives the maximum and minimum value of principal stress denoted by σ_1 and σ_2. This are express as

$$\sigma_1 = \frac{1}{2}(\sigma_{xx} + \sigma_{yy}) + \sqrt{\left(\frac{\sigma_{xx} - \sigma_{yy}}{2}\right)^2 + \tau_{xy}^2}$$

$$\sigma_2 = \frac{1}{2}(\sigma_{xx} + \sigma_{yy}) - \sqrt{\left(\frac{\sigma_{xx} - \sigma_{yy}}{2}\right)^2 + \tau_{xy}^2}$$

Take $\sigma_i = \sigma_1 / \sigma_2 \mid \tau_i = \tau_{xy}$, then squaring both equations and adding them, we obtain as:

$$\left(\sigma_i - \frac{\sigma_{xx} + \sigma_{yy}}{2}\right)^2 + \tau_i^2 = \left(\frac{\sigma_{xx} - \sigma_{yy}}{2}\right)^2$$

INTRODUCTION TO FLUID KINEMATICS

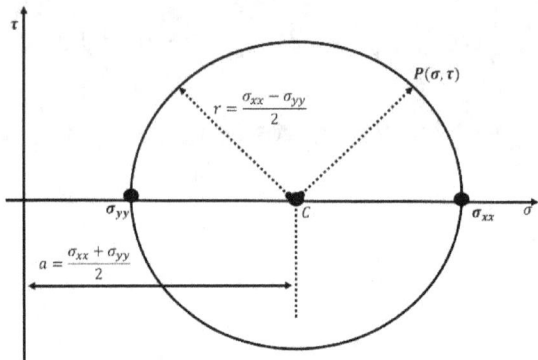

which represents equation of circle whose centre, $C\left(\dfrac{\sigma_{xx}+\sigma_{yy}}{2},0\right)$ and radius, $r=\dfrac{\sigma_{xx}-\sigma_{yy}}{2}$ as shown in figure. This circle known as Mohr's circle for a state of stress in two-dimensional coordinate system.

Mohr's circle for a state of stress in 3D coordinate system

The components of stress tensor along the principal planes at an arbitrary point P of a continuous medium is given by

$$\sigma_{ij} = \begin{pmatrix} \sigma_{xx} & 0 & 0 \\ 0 & \sigma_{yy} & 0 \\ 0 & 0 & \sigma_{zz} \end{pmatrix}$$

$$r = \sigma_{ij}.\hat{n} = \begin{pmatrix} \sigma_{xx} & 0 & 0 \\ 0 & \sigma_{yy} & 0 \\ 0 & 0 & \sigma_{zz} \end{pmatrix}\begin{pmatrix} n_1 \\ n_2 \\ n_3 \end{pmatrix} = \begin{pmatrix} \sigma_{xx}n_1 \\ \sigma_{yy}n_2 \\ \sigma_{zz}n_3 \end{pmatrix}$$

The normal component of stress is

$$\sigma = r^T.\hat{n} = \begin{pmatrix} \sigma_{xx}n_1 & \sigma_{yy}n_2 & \sigma_{zz}n_3 \end{pmatrix}\begin{pmatrix} n_1 \\ n_2 \\ n_3 \end{pmatrix} = \sigma_{xx}n_1^2 + \sigma_{yy}n_2^2 + \sigma_{zz}n_3^2 \quad \text{------(1)}$$

Since principal stress and tangential shear stress form a couple in a system. Thus, we have

INTRODUCTION TO FLUID KINEMATICS

$$\sigma_{xx}^2 n_1^2 + \sigma_{yy}^2 n_2^2 + \sigma_{zz}^2 n_3^2 = \sigma^2 + \tau^2 \quad \text{------(2)}$$

The unit normal vector $\hat{n}$ satisfies the condition $|\hat{n}| = 1 \Rightarrow n_1^2 + n_2^2 + n_3^2 = 1$ ---(3)

The system of equations given by (1), (2), and (3) can be written in matrix as

$$\begin{bmatrix} \sigma_{xx}^2 & \sigma_{yy}^2 & \sigma_{zz}^2 \\ \sigma_{xx} & \sigma_{yy} & \sigma_{zz} \\ 1 & 1 & 1 \end{bmatrix} \begin{bmatrix} n_1^2 \\ n_2^2 \\ n_3^2 \end{bmatrix} = \begin{bmatrix} \sigma^2 + \tau^2 \\ \sigma \\ 1 \end{bmatrix}$$

With constraints $0 \le n_1^2 \le 1, \ 0 \le n_2^2 \le 1, \ 0 \le n_3^2 \le 1$

The feasible solution for $x = \begin{bmatrix} n_1^2 & n_2^2 & n_3^2 \end{bmatrix}$ indicates the feasible point (σ, τ) on half space of feasible region.

The system of equations $Ax = b$ are rearrange as follow:

$$\sigma^2 + \tau^2 - (\sigma_{xx} + \sigma_{zz})\sigma + \sigma_{xx}\sigma_{zz} - \frac{A}{(\sigma_{xx} - \sigma_{zz})} n_1^2 = 0 \quad \text{----(4)}$$

$$\sigma^2 + \tau^2 - (\sigma_{yy} + \sigma_{zz})\sigma + \sigma_{yy}\sigma_{zz} - \frac{A}{(\sigma_{yy} - \sigma_{zz})} n_2^2 = 0 \quad \text{----(5)}$$

$$\sigma^2 + \tau^2 - (\sigma_{xx} + \sigma_{yy})\sigma + \sigma_{xx}\sigma_{yy} - \frac{A}{(\sigma_{xx} - \sigma_{yy})} n_3^2 = 0 \quad \text{-----(6)}$$

Where $A = (\sigma_{xx} - \sigma_{yy})(\sigma_{yy} - \sigma_{zz})(\sigma_{xx} - \sigma_{zz})$ Then equation (6) rewritten as

$$(\sigma - c_3)^2 + \tau^2 = r_3^2 \quad \text{------(7)}$$

which is equation of circle whose centre $c_3 \left[\frac{1}{2}(\sigma_{xx} + \sigma_{yy}), 0 \right]$

and radius $r_3 = \sqrt{\frac{1}{4}(\sigma_{xx} - \sigma_{yy})^2 + (\sigma_{yy} - \sigma_{zz})(\sigma_{xx} - \sigma_{zz})n_3^2}$.

INTRODUCTION TO FLUID KINEMATICS

Similarly, equation (5) and (4) rewritten as

$$(\sigma - c_2)^2 + \tau^2 = r_2^2 \quad \text{------(8)}$$

Which is equation of circle whose centre $c_2\left[\frac{1}{2}(\sigma_{xx} + \sigma_{zz}), 0\right]$ and

Radius $r_2 = \sqrt{\frac{1}{4}(\sigma_{xx} - \sigma_{zz})^2 + (\sigma_{xx} - \sigma_{yy})(\sigma_{zz} - \sigma_{yy})n_2^2}$.

$$(\sigma - c_1)^2 + \tau^2 = r_1^2 \quad \text{-----(9)}$$

Which is equation of circle whose centre $c_1\left[\frac{1}{2}(\sigma_{yy} + \sigma_{zz}), 0\right]$ and

Radius $r_1 = \sqrt{\frac{1}{4}(\sigma_{yy} - \sigma_{zz})^2 + (\sigma_{yy} - \sigma_{xx})(\sigma_{zz} - \sigma_{xx})n_1^2}$.

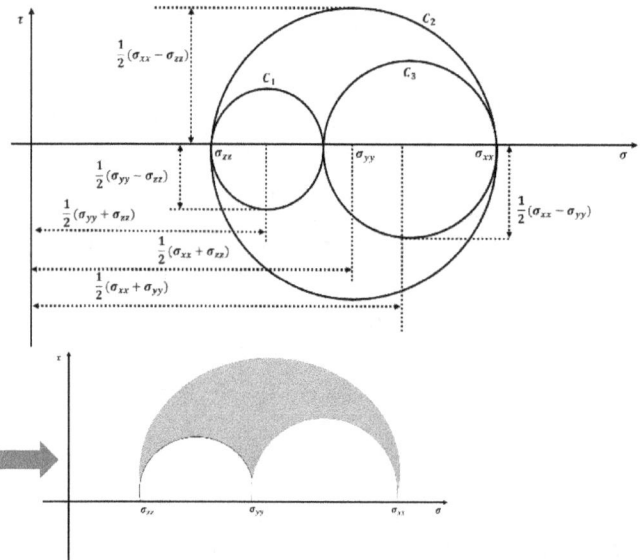

Thus, intersection of three circle given by equations (9), (8), and (7) determine the feasible region in the Mohr's circle as

shown in figure. Clearly, it has observed that every point of the feasible region in the Mohr's circle represents the state of stress at any particular point w.r.t respective planes.

BOOSTER CAPSULE : DRAG AND LIFT
INTRODUCTION

Flowing of air over buildings is an example of flowing of fluid over a stationary body where some forces exerted by the fluid on the body. Similarly, there are some forces exerted on the stationary body as we observed on flowing of water over bridges.

Movement of ships and submarines through water is an example of movement of bodies in a stationary fluid where some forces is exerted by the fluid on the body. Similar happening observed in the case of flying aeroplanes through air.

Moreover, these are some forces exerted by the fluid on the body when both body and fluid are moving at different speed.

1. FORCES EXERTED BY A FLOWING FLUID ON A STATIONARY BODY
2. FORCES EXERTED BY A STATIONARY FLUID ON A MOVING BODY
3. FORCES EXERTED BY A FLUID ON A BODY WHEN BOTH ARE IN MOVEMENT

FORCES EXERTED BY A FLOWING FLUID ON A STATIONARY BODY

Consider a static body in a real fluid, which is flowing at a uniform velocity v as shown in figure.

INTRODUCTION TO FLUID KINEMATICS

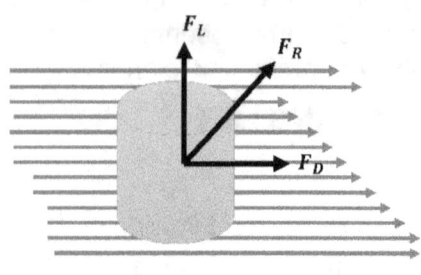

The flowing fluid exerted a force on the stationary body. The force (F_R) exerted by the flowing fluid on the static body. The force (F_R) is perpendicular to the surface of the body and also inclined to the direction of fluid motion. The total force (F_R) can be resolved in two components, one is along the direction of fluid motion and other one is perpendicular to the direction of fluid motion.

DRAG

The component of the force (F_R) along the direction of fluid motion is called drag. This component of force symbolize as F_D.

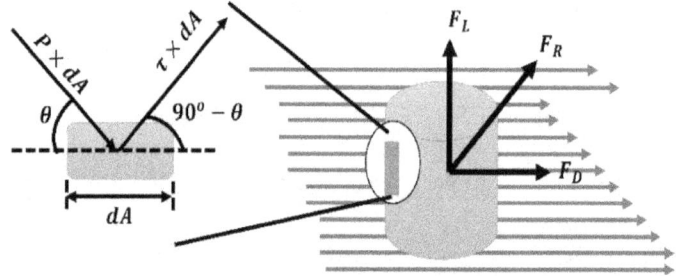

Consider an arbitrary shaped solid body placed in a real fluid, which is flowing with a uniform velocity v in a horizontal direction as shown in figure. Consider a small elemental area dA on the surface of the body. The forces on the surface area dA are:

1. Pressure $= P \times dA$, acting perpendicular to the free surface
2. Shear force $= \tau \times dA$, acting along the tangential direction to the free surface.

Let angle made by pressure with horizontal direction be θ.

INTRODUCTION TO FLUID KINEMATICS

Then drag force on elemental area

= Force due to pressure in the direction of fluid motion

+ Force due to shear stress in the direction of fluid motion.

$$= P \times dA \cos\theta + \tau dA \cos(90° - \theta)$$

$$= P \times dA \cos\theta + \tau\, dA \sin\theta$$

∴ Total drag, F_D

= summation of $P \times dA\cos\theta$ + summation of $\tau dA \sin\theta$

$$\Rightarrow F_D = \int P \cos\theta\, dA + \int \tau \sin\theta\, dA \quad ----(1)$$

where the terms called as

$$\int P \cos\theta\, dA \quad + \quad \int \tau \sin\theta\, dA$$

⬆ **Pressure Drag**   ⬆ **Friction Drag**

LIFT

The component of the total force (F_R) acting perpendicular to the direction of fluid motion is called Lift. This component denoted by F_L. Lift force occurs only when the axis of the body is inclined to the direction of fluid motion. The Lift force is zero when the axis of the body is parallel to the direction of fluid motion.

From the figure, we can calculate the Lift force on elemental area

= Force due to pressure acting perpendicular to the fluid

INTRODUCTION TO FLUID KINEMATICS

motion direction + Force due to shear stress acting perpendicular to fluid motion direction.

$$= -P \times dA \sin\theta + \tau \times dA \sin(90° - \theta)$$

$$= -P \times dA \sin\theta + \tau \, dA \cos\theta$$

$-ve$ sign is taken with pressure force as it is acting in the downward direction and $+ve$ sign is taken with shear force as it is acting in the upward direction.

∴ Total lift, F_L

= summation of $(-P \times dA \sin\theta)$ + summation of $\tau \, dA \sin\theta$

$$\Rightarrow F_L = \int \tau \cos\theta \, dA - \int P \sin\theta \, dA \quad ---(2)$$

The drag force and Lift force can be determine from the equation (1) and (2) when the pressure distribution and shear stress distribution along the surface of the body are given. However, determination of the level of pressure and shear distribution is a very tedious task which makes difficult to calculate the drag force and Lift force. Thus, an alternative to determine the values of Drag force and Lift force with help of numerical technique.

The Drag force and a Lift force for a static body submerged in moving fluid of density ρ, at a uniform velocity v can be calculate as

$$F_D = C_D \, \rho \, A \frac{v^2}{2} \quad \text{and} \quad F_L = C_L \, \rho \, A \frac{v^2}{2}$$

where C_D = Co-efficient of drag

C_L = Co-efficient of Lift

A = area of immersed body

Hence, resultant force on the body is given by

$$F_R = \sqrt{F_D^2 + F_L^2}$$

HIGHER ORDER THINKING SKILL QUESTION

1. **Calculate the viscous stress induced in Couette Flow.**

```
Upper plate is moving along x-axis
h  Fluid of density ρ and viscosity μ
Lower plate is fixed in position
```

Consider steady, incompressible, laminar flow of a Newtonian fluid past through two infinite parallel plates as shown in figure. In this case, upper plate is moving with constant speed $\vec{q} = u\hat{i} + v\hat{j} + w\hat{k}$, and lower plate is fixed in its position. The two plates are h distance apart vertically. Here, there is no external force are being applied but hydrostatic pressure induced on the flow due to gravity. So, force due to gravity acts in the negative $z-$ direction which cannot be depict in figure. The flow with set of given conditions are known as Couette flow. To find out the viscous stress induced in Couette Flow, we need to apply following assumptions to the governing equations of fluid flow.

1. The plates are infinite in length in x-direction.
2. The flow is steady (i.e., $\dfrac{\partial \vec{q}}{\partial t} = 0$)
3. The fluid flow is parallel to x-axis.
4. There is no pressure gradient acts on fluid in x-direction.

INTRODUCTION TO FLUID KINEMATICS

5. The velocity field is two-dimensional, $\left(w=0 \mid \dfrac{\partial \vec{q}}{\partial z}=0\right)$

6. Gravity acts in the negative z-direction $\left(g_x = g_y = 0 \mid g_z = -g \Rightarrow \vec{g} = -g\hat{k}\right)$

The continuity equation in cartesian coordinates for two-dimensional flow is

$$\frac{\partial u}{\partial x} + \frac{\partial v}{\partial y} = 0 \quad \text{------(1)}$$

On applying assumption 3 (*flow is parallel to x-axis*) means that y-component of fluid velocity is zero,

$\left(v = 0 \Rightarrow \dfrac{\partial v}{\partial y} = 0\right)$. Then equation (1) reduced to

$$\frac{\partial u}{\partial x} = 0 \quad \text{-----(II)}$$

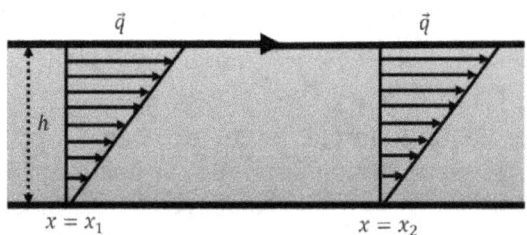

The above expression indicates that u is not function of x. It means that the velocity profile of the flow is same at any position on x-axis. Due to this reason the Couette flow considered as fully developed flow. Since u is not function of time t (assumption 2) and u is not a function of z (assumption 5), so the only possibility is that u is a function of y, and it can be express as

$$u = u(y) \quad \text{-------(3)}$$

In this case, consider x-component of Navier-Stokes equation to determine the fluid motion as

INTRODUCTION TO FLUID KINEMATICS

$$\rho\left(\frac{\partial u}{\partial t}+u\frac{\partial u}{\partial x}+v\frac{\partial u}{\partial y}+w\frac{\partial u}{\partial z}\right)$$

$$=-\frac{\partial P}{\partial x}+\rho g_x+\mu\left(\frac{\partial^2 u}{\partial x^2}+\frac{\partial^2 u}{\partial y^2}+\frac{\partial^2 u}{\partial z^2}\right) \quad \text{----(4)}$$

Since $\frac{\partial u}{\partial x}=0 \Rightarrow \frac{\partial^2 u}{\partial x^2}=0$ and $u\frac{\partial u}{\partial x}=0$

Since the fluid flow is steady $\Rightarrow \frac{\partial u}{\partial t}=0$

Since y-component of the velocity is zero, $v=0 \Rightarrow v\frac{\partial u}{\partial y}=0$

Since the flow is two-dimensional, $w=0 \Rightarrow w\frac{\partial u}{\partial z}=0$ and $\frac{\partial^2 u}{\partial z^2}=0$

Since there is no external pressure force acts along x-direction $\Rightarrow \frac{\partial P}{\partial x}=0$

Since gravity acts in the negative z-direction $\Rightarrow \rho g_x=0$

Using above facts, the equation (4) reduced to

$$\frac{d^2 u}{dy^2}=0$$

Integrating u w.r.t y, we get

$$\Rightarrow u = k_1 y + k_2 \quad \text{------(5)}$$

Boundary conditions:
On upper plate: $y=h, u=q, v=0, w=0$
Om lower plate: $y=0, u=v=w=0$

Applying boundary conditions, we obtain the value of k_1 and k_2 as

INTRODUCTION TO FLUID KINEMATICS

$$(5) \Rightarrow 0 = k_1 \times 0 + k_2 \Rightarrow k_2 = 0$$

$$(5) \Rightarrow q = k_1 \times h + 0 \Rightarrow k_1 = \frac{q}{h}$$

Put the value of k_1 and k_2 in equation (5), we get

$$u = \frac{q}{h} y \quad \text{-----(6)}$$

To calculate the shear stress acts on lower plate, we consider a infinitesimal fluid element whose bottom face is in contact with the lower plate as shown in figure. Mathematically positive viscous stress are shown in figure. In the Couette flow, the stresses are in proper direction as fluid above the element pull it to the right while the boundary of lower plate below the element pulls it to the left. Mathematically, the viscous stress tensor can be express as

$$\tau_{ij} = \begin{bmatrix} 2\mu \frac{\partial u}{\partial x} & \mu\left(\frac{\partial u}{\partial y} + \frac{\partial v}{\partial x}\right) & \mu\left(\frac{\partial u}{\partial z} + \frac{\partial w}{\partial x}\right) \\ \mu\left(\frac{\partial u}{\partial y} + \frac{\partial v}{\partial x}\right) & 2\mu \frac{\partial v}{\partial y} & \mu\left(\frac{\partial v}{\partial z} + \frac{\partial w}{\partial y}\right) \\ \mu\left(\frac{\partial w}{\partial x} + \frac{\partial u}{\partial z}\right) & \mu\left(\frac{\partial w}{\partial y} + \frac{\partial v}{\partial z}\right) & 2\mu \frac{\partial w}{\partial z} \end{bmatrix}$$

$$\Rightarrow \tau_{ij} = \begin{bmatrix} 0 & \mu \frac{q}{h} & 0 \\ \mu \frac{q}{h} & 0 & 0 \\ 0 & 0 & 0 \end{bmatrix}$$

INTRODUCTION TO FLUID KINEMATICS

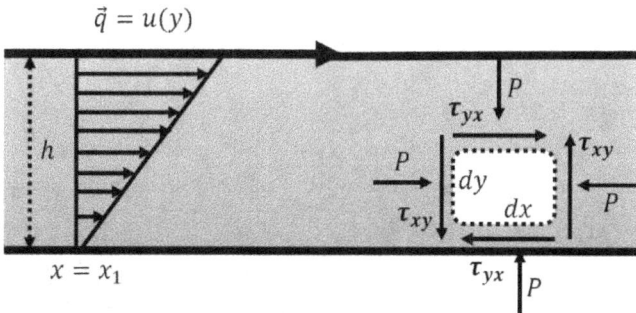

The shear force per unit area on the boundary of lower plate is equal to $\tau_{yx} = \mu \dfrac{q}{h}$ and acts in the negative x-direction. The shear force per unit area on the surface of the plates is equal and opposite to this. So, shear force per unit area acting on the boundary is defined as

$$\frac{\vec{F}}{A} = \mu \frac{q}{h} \hat{i}$$

2. Calculate the viscous stress induced in Haigen Flow

Consider steady, incompressible, laminar flow of a Newtonian fluid past through infinitely circular pipe of radius R as shown in figure. In this case, the fluid flow is parallel to wall of the circular pipe. The two plates are $D = 2R$ distance apart vertically. Here, there is no external force are being applied but hydrostatic pressure induced on the flow due to pressure gradient. So, force due to gravity less significant. The flow with set of given conditions are known as Haigen Poiseuille flow.

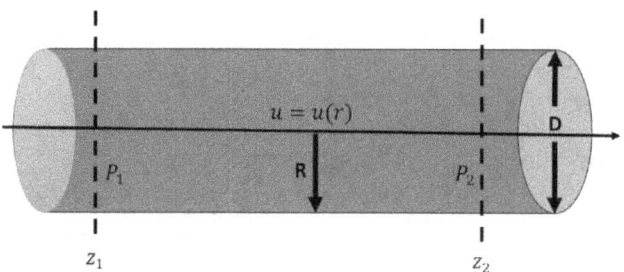

INTRODUCTION TO FLUID KINEMATICS

To find out the viscous stress induced in Haigen Poiseuille Flow, we need to apply following assumptions to the governing equations of fluid flow.

1. The circular pipe is infinitely long in $z-$ direction.
2. The fluid is incompressible which indicates that density, $\rho =$ constant.
3. The fluid flow is steady with respect to the time, it implies that all partial derivatives w.r.t. instant of time are zero.
4. The fluid is Newtonian fluid and flow is laminar as shown in figure.
5. In order to find out the expression of viscous stress of Haigen Poiseuille flow, we apply cylindrical coordinate system on a control volume of a fluid element closet to the wall of the circular pipe. For simplification in calculation, we apply cylindrical coordinate (r,θ,z) system to velocity field which converts to (u_r, u_θ, u_z),
6. The fluid flow is parallel to $z-$ direction as a result the radial component of velocity field is zero $(u_r = 0)$
7. The fluid flow is axisymmetric without any revolution which indicates that angular component of velocity field is zero $(u_\theta = 0)$.
8. Assume that there is no gravitational force acts on flow.
9. As the fluid is incompressible with constant pressure gradient acts on $z-$ direction.

$$\frac{\partial P}{\partial z} = \frac{p_2 - p_1}{z_2 - z_1}$$

10. The no-slip conditions applied on the flow.

The continuity equation in cylindrical coordinate system is given by

INTRODUCTION TO FLUID KINEMATICS

$$\frac{\partial \rho}{\partial t} + \left[\frac{1}{r}\frac{\partial}{\partial r}(\rho r u_r) + \frac{1}{r}\frac{\partial}{\partial \theta}(\rho u_\theta) + \frac{\partial}{\partial z}(\rho u_z)\right] = 0 ---(1)$$

According to the assumption 3, we have $\frac{\partial P}{\partial t} = 0$ Then equ (1) reduced to

$$\frac{1}{r}\frac{\partial}{\partial r}(\rho r u_r) + \frac{1}{r}\frac{\partial}{\partial \theta}(\rho u_\theta) + \frac{\partial}{\partial z}(\rho u_z) = 0 ---(2)$$

According to the assumption 2, we have ρ = constant. Then equ (2) reduced to

$$\rho\left[\frac{1}{r}\frac{\partial}{\partial r}(r u_r) + \frac{1}{r}\frac{\partial}{\partial \theta}(u_\theta) + \frac{\partial}{\partial z}(u_z)\right] = 0$$

$$\Rightarrow \frac{1}{r}\frac{\partial}{\partial r}(r u_r) + \frac{1}{r}\frac{\partial}{\partial \theta}(u_\theta) + \frac{\partial}{\partial z}(u_z) = 0 ---(3)$$

According to the assumption 6, we have $u_r = 0$. Then equ (3) reduced to

$$\frac{1}{r}\frac{\partial}{\partial \theta}(u_\theta) + \frac{\partial}{\partial z}(u_z) = 0 ---(4)$$

According to the assumption 7, we have $u_\theta = 0$. Then equ (4) reduced to

$$\frac{\partial}{\partial z}(u_z) = 0 ---(5)$$

The above expression indicates that u is not function of z. It means that the velocity profile of the flow is same at any position on $z-$ axis. Due to this reason the Haigen Poiseuille flow considered as fully developed flow. Since u is not function of time t (assumption 3) and u is not a function of θ (assumption 7), so the only possibility is that u is a function of r, and it can be express as

INTRODUCTION TO FLUID KINEMATICS

$$u = u(r) \text{ only} ----- (6)$$

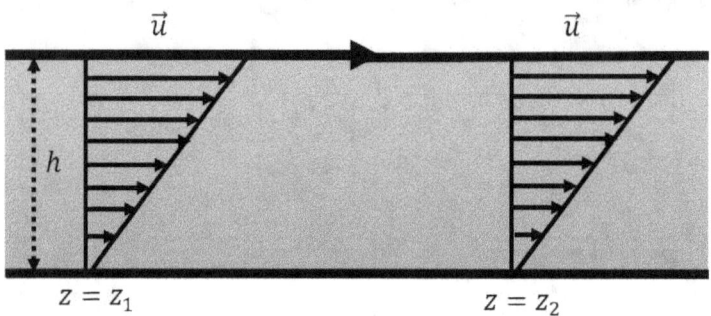

The Navier Strokes equation in axial direction (in cylindrical coordinate system) is given by

$$\rho g_z - \frac{\partial P}{\partial z} + \frac{\partial}{\partial r}\left[\mu\left(\frac{\partial u_r}{\partial z} + \frac{\partial u_z}{\partial r}\right)\right] + \frac{1}{r}\frac{\partial}{\partial \theta}\left[\mu\left(\frac{1}{r}\frac{\partial u_z}{\partial \theta} + \frac{\partial u_\theta}{\partial z}\right)\right]$$

$$+ \frac{\partial}{\partial z}\left[\mu\left(2\frac{\partial u_z}{\partial z} - \frac{2}{3}\nabla.u\right)\right] + \frac{\mu}{r}\left(\frac{\partial u_r}{\partial z} + \frac{\partial u_z}{\partial r}\right) = \rho\left[\frac{Du}{Dt}\right]$$

where $\dfrac{D}{Dt} = \dfrac{\partial}{\partial t} + u_r \dfrac{\partial}{\partial r} + \dfrac{u_\theta}{r}\dfrac{\partial}{\partial \theta} + u_z \dfrac{\partial}{\partial z}$

$----- \rightarrow (A)$

Right Hand Side of Equation (A) reduced to

$$\rho\left[\frac{Du}{Dt}\right] = \rho\left[\frac{\partial u}{\partial t} + u_r \frac{\partial u}{\partial r} + \frac{u_\theta}{r}\frac{\partial u}{\partial \theta} + u_z \frac{\partial u}{\partial z}\right] --(7)$$

According to the assumption 3, we have $\dfrac{\partial u}{\partial t} = 0$ Then equ (7) reduced to

$$\rho\left[\frac{Du}{Dt}\right] = \rho\left[u_r \frac{\partial u}{\partial r} + \frac{u_\theta}{r}\frac{\partial u}{\partial \theta} + u_z \frac{\partial u}{\partial z}\right] --(8)$$

INTRODUCTION TO FLUID KINEMATICS

According to the assumption 6, we have $u_r = 0$. Then equ (8) reduced to

$$\rho\left[\frac{Du}{Dt}\right] = \rho\left[\frac{u_\theta}{r}\frac{\partial u}{\partial \theta} + u_z\frac{\partial u}{\partial z}\right] \quad --(9)$$

According to the assumption 7, we have $u_\theta = 0$. Then equ (9) reduced to

$$\rho\left[\frac{Du}{Dt}\right] = \rho\left[u_z\frac{\partial u}{\partial z}\right] \quad ------(10)$$

From the continuity equation of the given flow, we derived that $\frac{\partial u_z}{\partial z} = 0$. Then equ (10) reduced to

$$\rho\left[\frac{Du}{Dt}\right] = 0 \quad ------(11)$$

Left Hand side of the equ (A) can be expressed as

$$\rho g_z - \frac{\partial P}{\partial z} + \frac{\partial}{\partial r}\left[\mu\left(\frac{\partial u_r}{\partial z} + \frac{\partial u_z}{\partial r}\right)\right] + \frac{1}{r}\frac{\partial}{\partial \theta}\left[\mu\left(\frac{1}{r}\frac{\partial u_z}{\partial \theta} + \frac{\partial u_\theta}{\partial z}\right)\right]$$
$$+ \frac{\partial}{\partial z}\left[\mu\left(2\frac{\partial u_z}{\partial z} - \frac{2}{3}\nabla.u\right)\right] + \frac{\mu}{r}\left(\frac{\partial u_r}{\partial z} + \frac{\partial u_z}{\partial r}\right) \quad -----(12)$$

According to the assumption 8, we have $g_z = 0$. Then equ (12) reduced to

$$-\frac{\partial P}{\partial z} + \frac{\partial}{\partial r}\left[\mu\left(\frac{\partial u_r}{\partial z} + \frac{\partial u_z}{\partial r}\right)\right] + \frac{1}{r}\frac{\partial}{\partial \theta}\left[\mu\left(\frac{1}{r}\frac{\partial u_z}{\partial \theta} + \frac{\partial u_\theta}{\partial z}\right)\right]$$
$$+ \frac{\partial}{\partial z}\left[\mu\left(2\frac{\partial u_z}{\partial z} - \frac{2}{3}\nabla.u\right)\right] + \frac{\mu}{r}\left(\frac{\partial u_r}{\partial z} + \frac{\partial u_z}{\partial r}\right) \quad -----(13)$$

INTRODUCTION TO FLUID KINEMATICS

According to the assumption 7, we have $u_\theta = 0$. Then equ (13) reduced to

$$-\frac{\partial P}{\partial z} + \frac{\partial}{\partial r}\left[\mu\left(\frac{\partial u_r}{\partial z} + \frac{\partial u_z}{\partial r}\right)\right] + 0$$

$$+ \frac{\partial}{\partial z}\left[\mu\left(2\frac{\partial u_z}{\partial z} - \frac{2}{3}\nabla . u\right)\right] + \frac{\mu}{r}\left(\frac{\partial u_r}{\partial z} + \frac{\partial u_z}{\partial r}\right) \quad \text{-----(14)}$$

From the continuity equation of the given flow, we derived that $\frac{\partial u_z}{\partial z} = 0$. Then equ (10) reduced to

$$-\frac{\partial P}{\partial z} + \frac{\partial}{\partial r}\left[\mu\left(\frac{\partial u_r}{\partial z} + \frac{\partial u_z}{\partial r}\right)\right] + 0$$

$$+ 0 + \frac{\mu}{r}\left(\frac{\partial u_r}{\partial z} + \frac{\partial u_z}{\partial r}\right) \quad \text{-----(15)}$$

From the continuity equation of the given flow, we derived that u is a function of r. Then equ (15) reduced to

$$-\frac{\partial P}{\partial z} + \frac{\partial}{\partial r}\left[\mu\left(0 + \frac{\partial u}{\partial r}\right)\right] + \frac{\mu}{r}\left(0 + \frac{\partial u}{\partial r}\right) \quad \text{-----(16)}$$

$$\Rightarrow -\frac{\partial P}{\partial z} + \frac{\partial}{\partial r}\left[\mu\left(0 + \frac{\partial u}{\partial r}\right)\right] + \frac{\mu}{r}\left(0 + \frac{\partial u}{\partial r}\right)$$

$$\Rightarrow -\frac{\partial P}{\partial z} + \frac{\mu}{r}\frac{\partial}{\partial r}\left(r\frac{\partial u}{\partial r}\right) \quad \text{-----(17)}$$

Combining equ (11) and equ (17), we express equ (A) as

INTRODUCTION TO FLUID KINEMATICS

$$-\frac{\partial P}{\partial z} + \frac{\mu}{r}\frac{\partial}{\partial r}\left(r\frac{\partial u}{\partial r}\right) = 0$$

$$\Rightarrow \frac{1}{r}\frac{\partial}{\partial r}\left(r\frac{\partial u}{\partial r}\right) = \frac{1}{\mu}\frac{\partial P}{\partial z} \quad\text{-----(18)}$$

In order to find out the velocity field of the given flow, we need to convert the partial differential equation (18) into complete differential equation as

$$\frac{1}{r}\frac{d}{dr}\left(r\frac{du}{dr}\right) = \frac{1}{\mu}\frac{dP}{dz} \quad\text{-----(19)}$$

Integrate the equ (19) w.r.t. 'r', we get

$$r\frac{du}{dr} = \frac{r^2}{2\mu}\frac{dP}{dz} + C_1 \quad\text{-----(20)}$$

Where C_1 is constant of integration.

Again, integrate the equ (20) w.r.t. 'r', we get

$$u = \frac{r^2}{4\mu}\frac{dP}{dz} + C_1 \log_e(r) + C_2 \quad\text{-----(21)}$$

Where C_1 and C_2 are constant of integration.

According to the assumption 10, we can apply no-slip condition (i.e., $\frac{du}{dr} = 0$ when $r = 0$) on the equ (20) gives

$$0 = 0 + C_1 \Rightarrow C_1 = 0$$

Since, the given flow is fully developed flow. So, we can apply boundary condition BC: $u(r) = 0$ when $r = R$ on the

INTRODUCTION TO FLUID KINEMATICS

equ (21) which gives $0 = \dfrac{R^2}{4\mu}\dfrac{dP}{dx} + C_2 \Rightarrow C_2 = -\dfrac{R^2}{4\mu}\dfrac{dP}{dz}$

Next, the equ (21) reduced to

$$u = \dfrac{r^2}{4\mu}\dfrac{dP}{dz} - \dfrac{R^2}{4\mu}\dfrac{dP}{dz}$$

$$\Rightarrow u = \dfrac{1}{4\mu}\dfrac{dP}{dz}(r^2 - R^2) \;----(22)$$

To calculate the viscous shear stress acts on wall of circular pipe, we consider a infinitesimal fluid element whose bottom face is in contact with wall of circular pipe as shown in figure. Mathematically positive viscous stress are shown in figure. The stress tensor in cylindrical coordinate expressed as

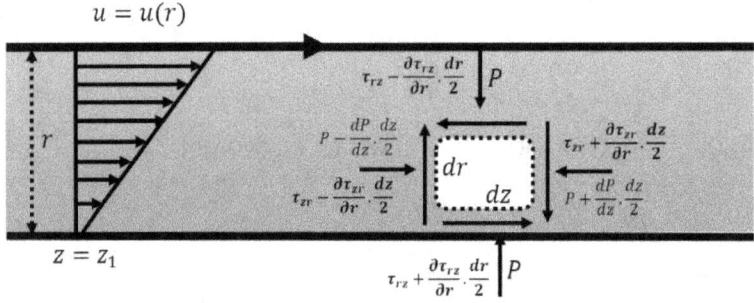

$$\tau = \begin{bmatrix} \tau_{rr} & \tau_{r\theta} & \tau_{rz} \\ \tau_{\theta r} & \tau_{\theta\theta} & \tau_{\theta z} \\ \tau_{zr} & \tau_{z\theta} & \tau_{zz} \end{bmatrix} = \begin{bmatrix} 0 & 0 & \mu\dfrac{\partial u}{\partial r} \\ 0 & 0 & 0 \\ \mu\dfrac{\partial u}{\partial r} & 0 & 0 \end{bmatrix} \;--(23)$$

We can derived that the viscous shear stress at the wall of circular pipe is given by

$$\tau_{rz} = \mu\dfrac{\partial u}{\partial r} = \dfrac{r}{2}\dfrac{dP}{dz}$$

INTRODUCTION TO FLUID KINEMATICS

As the given fluid flows from left to right direction through circular pipe, as a result pressure gradient $\left(\dfrac{dP}{dz}\right)$ is negative and viscous shear stress on the bottom of the fluid element is in the opposite direction as shown in figure. The viscous shear force per unit area acts on the wall is given by

$$\dfrac{\vec{F}}{A} = -\dfrac{r}{2}\dfrac{dP}{dz}\vec{i}$$

4. STRAIN

INTRODUCTION

In order to determine the forces acts on the fluid element, the study of rate of deformation or **strain** is important in relation with different fluid flows. A fluid is continuously deforms, and the parameter to measure the rate of deformation in flows considered as **strain**. The deformation of a fluid element is similar to that happen in the case of solids, which defines as normal strain. **Normal strain** is defined as the change in length per unit of a fluid element in horizontal direction. **Shear strain** is defined as the change in shape per unit of a fluid element in vertical direction.

NORMAL STRAIN RATE

Consider a fluid element positioned at AB deforms to the new position A'B' due to applied force in the x-direction as shown in figure.

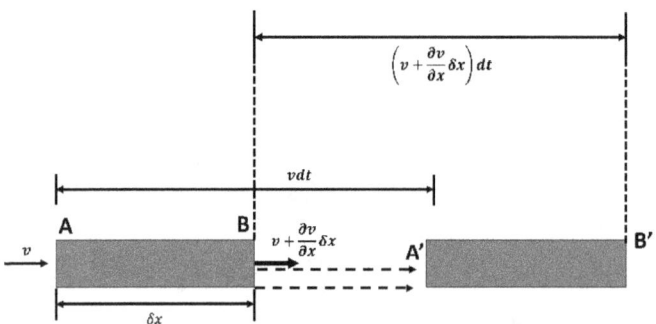

The rate of change of length per unit is defined as

INTRODUCTION TO FLUID KINEMATICS

$$\frac{1}{\delta x}\frac{D}{Dt}(\delta x) = \frac{1}{dt}\left[\frac{A'B' - AB}{AB}\right]$$

$$\Rightarrow \frac{1}{\delta x}\frac{D}{Dt}(\delta x) = \frac{1}{dt}\left[\frac{\delta x + \frac{\partial v}{\partial x}\delta x dt - \delta x}{\delta x}\right]$$

$$\Rightarrow \frac{1}{\delta x}\frac{D}{Dt}(\delta x) = \frac{\partial v}{\partial x}$$

The material derivative D/Dt is used as we have implicitly followed a fluid element. In general, the normal strain rate in the i direction is defined as

$$\frac{1}{\delta x_i}\frac{D}{Dt}(\delta x_i) = \frac{\partial v_i}{\partial x_i}$$

SHEAR STRAIN RATE

Subsequently, a fluid element deform in shape. The shear strain rate of a fluid element is defined as the rate of increase of the angle formed by two mutually perpendicular axes. Consider a fluid element positioned with sides parallel to the coordinate axes at the time t, and undergoing strain the element changes its shape at the time $t + dt$ as shown in figure.

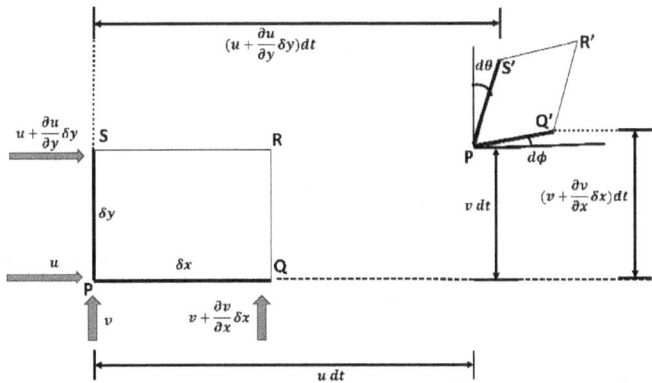

The rate of change of dimension is defined as

$$\frac{d\theta + d\phi}{dt} = \frac{1}{dt}\left[\frac{1}{\delta y}\left(\frac{\partial u}{\partial y}\delta y dt\right) + \frac{1}{\delta x}\left(\frac{\partial v}{\partial x}\delta x dt\right)\right]$$

$$\Rightarrow \frac{d\theta + d\phi}{dt} = \frac{\partial u}{\partial y} + \frac{\partial v}{\partial x}$$

In general, the shear strain rate in i,j direction can be expressed as

$$\frac{d\theta + d\phi}{dt} = \frac{\partial v_i}{\partial x_j} + \frac{\partial v_j}{\partial x_i}$$

VOLUMETRIC STRAIN RATE

The sum of the normal strain rates in the three mutually orthogonal directions is the rate of change of volume per unit volume, known as **volumetric strain rate**. Consider a fluid element of sides $\delta x_1, \delta x_2,$ and δx_3 whose volume δV defined as

$$\delta V = \delta x_1 \delta x_2 \delta x_3$$

Then volumetric strain rate is given by

$$\frac{1}{\delta V}\frac{D}{Dt}(\delta V) = \frac{1}{\delta x_1 \delta x_2 \delta x_3}\frac{D}{Dt}(\delta x_1 \delta x_2 \delta x_3)$$

$$\Rightarrow \frac{1}{\delta V}\frac{D}{Dt}(\delta V) = \frac{1}{\delta x_1}\frac{D}{Dt}(\delta x_1) + \frac{1}{\delta x_2}\frac{D}{Dt}(\delta x_2) + \frac{1}{\delta x_3}\frac{D}{Dt}(\delta x_3)$$

$$\Rightarrow \frac{1}{\delta V}\frac{D}{Dt}(\delta V) = \frac{\partial v_1}{\partial x_1} + \frac{\partial v_2}{\partial x_2} + \frac{\partial v_3}{\partial x_3} = \sum_{i=1}^{3}\frac{\partial v_i}{\partial x_i}$$

GENERAL ANALYSIS OF FLUID MOTION

When the fluid element moves in a streamflow, it undergoes various motion. The following motion are briefly discussed:

Translation

It is the situation where a fluid element moves without changing its configuration. The movement of a rectangular fluid element from $ABCD$ to the new position $A'B'C'D'$ by translatory motion as shown in figure.

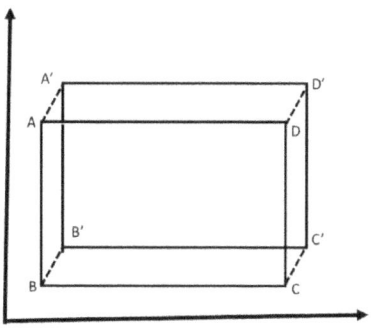

Rotation

It is the situation where a fluid element rotates about a certain axis without change in its configuration. The movement of a rectangular fluid element from *ABCD* to the new position *A'BC'D'* by rotatory motion as shown in figure.

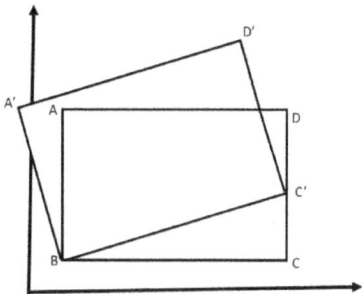

Linear Deformation

It is the situation where a fluid element changes its size but its position, shape unaltered. The movement of a rectangular fluid element from ABCD to the new position ABC'D' by translatory motion as shown in figure.

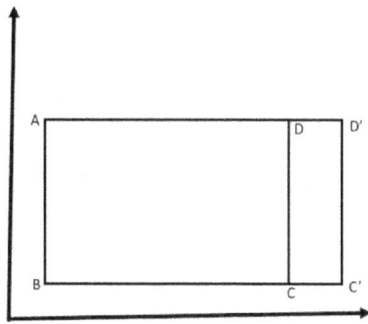

INTRODUCTION TO FLUID KINEMATICS

Angular Deformation

It is the situation where a fluid element change its shape by shearing strain but its size and position unaltered. The movement of a rectangular fluid element from ABCD to the new position A'BC'D' by translatory motion as shown in figure.

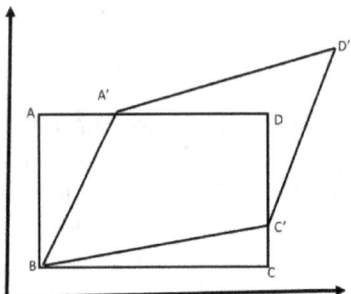

EXPRESSION FOR STRAIN TENSOR

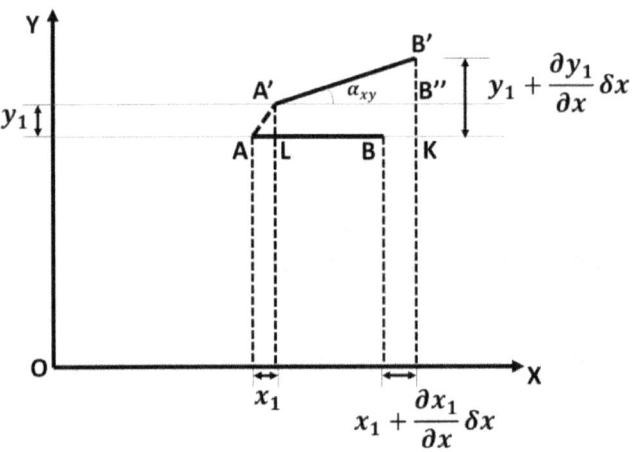

Before strain, the length of the edges are $AB = \delta x$

After very small strain, let AB occupy new position $A'B'$.

Let coordinates of A be (x,y) and that of A' be $(x + x_1, y + y_1)$

From the figure, the displacement of A along the axis OX is

138

INTRODUCTION TO FLUID KINEMATICS

x_1 i.e., $AL = x_1$, and along the axis OY is y_1 i.e., $A'L = y_1$

Next, the displacement of the point B along the x – axis is BK which is given by

$$BK = x_1 + \frac{\partial x_1}{\partial x} \delta x$$

Similarly, the displacement of the point B along the y – axis is $B'K$ which is given by

$$B'K = y_1 + \frac{\partial y_1}{\partial x} \delta x$$

Here, $A'B' = A'B''$ as $\angle B'A'B''$ is very small

The expression for $A'B'$ rewritten as

$$A'B' = A'B'' = (AB + BK) - AL$$

$$\Rightarrow A'B' = \delta x + x_1 + \frac{\partial x_1}{\partial x} \delta x - x_1$$

$$\Rightarrow A'B' = \delta x + \frac{\partial x_1}{\partial x} \delta x$$

The normal strain component in the x-direction as

$$\epsilon_{xx} = \frac{A'B' - AB}{AB}$$

$$\Rightarrow \epsilon_{xx} = \frac{\delta x + \frac{\partial x_1}{\partial x} \delta x - \delta x}{\delta x} = \frac{\frac{\partial x_1}{\partial x} \delta x}{\delta x}$$

$$\Rightarrow \epsilon_{xx} = \frac{\partial x_1}{\partial x}$$

Similarly, the normal strain component in the y-direction

$$\epsilon_{yy} = \frac{\partial y_1}{\partial y}$$

The expression for $B'B''$ rewritten as

$$B'B'' = B'K - B''K$$

$$\Rightarrow B'B'' = y_1 + \frac{\partial y_1}{\partial x} \delta x - y_1 = \frac{\partial y_1}{\partial x} \delta x$$

The shearing strain γ_{xy} at the point A is the change of angle between AB and AC.

The angle of rotation α_{yx} is given by

$$\alpha_{yx} \approx \tan \alpha_{yx} = \frac{B'B''}{A'B''}$$

$$\Rightarrow \alpha_{yx} = \frac{B'K - B''K}{A'B''} = \frac{B'B''}{A'B'}$$

$$\Rightarrow \alpha_{yx} = \frac{\frac{\partial y_1}{\partial x}\delta x}{\delta x + \frac{\partial x_1}{\partial x}\delta x} = \frac{\frac{\partial y_1}{\partial x}}{1 + \frac{\partial x_1}{\partial x}}$$

Here, we consider a very small strain, the quantity $\epsilon_{xx} = \frac{\partial x_1}{\partial x}$ ignore in the above fraction.

Then, we have

$$\alpha_{yx} = \frac{\partial y_1}{\partial x}$$

Similarly, considering the angle of rotation of the edge AC,

$$\alpha_{xy} = \frac{\partial x_1}{\partial y}$$

Then shearing strain at the point A is given by

$$\gamma_{xy} = \alpha_{yx} + \alpha_{xy}$$

$$\Rightarrow \gamma_{xy} = \frac{\partial y_1}{\partial x} + \frac{\partial x_1}{\partial y}$$

Similarly, we obtain the expressions for the normal strains and shearing strains in the other two coordinate planes.

If x_1, y_1 and z_1 be the components of displacement of a point in three-dimensional coordinates system, we obtain as:

Normal strains

$$\epsilon_{xx} = \frac{\partial x_1}{\partial x}, \quad \epsilon_{yy} = \frac{\partial y_1}{\partial y}, \quad \text{and} \quad \epsilon_{zz} = \frac{\partial z_1}{\partial z}$$

Shearing Strains

$$\gamma_{xy} = \gamma_{yx} = \frac{\partial y_1}{\partial x} + \frac{\partial x_1}{\partial y}$$

$$\gamma_{yz} = \gamma_{zy} = \frac{\partial z_1}{\partial y} + \frac{\partial y_1}{\partial z}$$

$$\gamma_{zx} = \gamma_{xz} = \frac{\partial x_1}{\partial z} + \frac{\partial z_1}{\partial x}$$

Note: The components of the shearing strains are expressed as

$$\gamma_{xy} = \gamma_{yx}, \quad \gamma_{yz} = \gamma_{zy}, \quad \gamma_{zx} = \gamma_{xz}$$

It follows that the shearing strains components are symmetric in their respective coordinates

RATE OF STRAIN

In fluid dynamics, we are concerns with the **rate of strain.**

Let u, v, and w be the component of the velocity of the fluid particle in the $x, y,$ and z direction respectively. Then, the components of rate of strain are given by

$$\epsilon_{xx} = \frac{\partial}{\partial t}\left(\frac{\partial x_1}{\partial x}\right) = \frac{\partial}{\partial x}\left(\frac{\partial x_1}{\partial t}\right) = \frac{\partial u}{\partial x}$$

$$\epsilon_{yy} = \frac{\partial}{\partial t}\left(\frac{\partial y_1}{\partial y}\right) = \frac{\partial}{\partial y}\left(\frac{\partial y_1}{\partial t}\right) = \frac{\partial v}{\partial y}$$

$$\epsilon_{zz} = \frac{\partial}{\partial t}\left(\frac{\partial z_1}{\partial z}\right) = \frac{\partial}{\partial z}\left(\frac{\partial z_1}{\partial t}\right) = \frac{\partial w}{\partial z}$$

$$\gamma_{xy} = \frac{\partial}{\partial t}\left(\frac{\partial y_1}{\partial x} + \frac{\partial x_1}{\partial y}\right) \Rightarrow \gamma_{xy} = \frac{\partial}{\partial t}\left(\frac{\partial y_1}{\partial x}\right) + \frac{\partial}{\partial t}\left(\frac{\partial x_1}{\partial y}\right) \Rightarrow \gamma_{xy}$$

$$= \frac{\partial}{\partial x}\left(\frac{\partial y_1}{\partial t}\right) + \frac{\partial}{\partial y}\left(\frac{\partial x_1}{\partial t}\right)$$

$$\Rightarrow \gamma_{xy} = \frac{\partial v}{\partial x} + \frac{\partial u}{\partial y} = \gamma_{yx}$$

INTRODUCTION TO FLUID KINEMATICS

$$\gamma_{yz} = \frac{\partial}{\partial t}\left(\frac{\partial z_1}{\partial y} + \frac{\partial y_1}{\partial z}\right) \Rightarrow \gamma_{yz} = \frac{\partial}{\partial t}\left(\frac{\partial z_1}{\partial y}\right) + \frac{\partial}{\partial t}\left(\frac{\partial y_1}{\partial z}\right) \Rightarrow \gamma_{yz}$$

$$= \frac{\partial}{\partial y}\left(\frac{\partial z_1}{\partial t}\right) + \frac{\partial}{\partial z}\left(\frac{\partial y_1}{\partial t}\right)$$

$$\Rightarrow \gamma_{yz} = \frac{\partial w}{\partial y} + \frac{\partial v}{\partial z} = \gamma_{zy}$$

$$\gamma_{zx} = \frac{\partial}{\partial t}\left(\frac{\partial x_1}{\partial z} + \frac{\partial z_1}{\partial x}\right) \Rightarrow \gamma_{zx} = \frac{\partial}{\partial t}\left(\frac{\partial x_1}{\partial z}\right) + \frac{\partial}{\partial t}\left(\frac{\partial z_1}{\partial x}\right) \Rightarrow \gamma_{zx}$$

$$= \frac{\partial}{\partial z}\left(\frac{\partial x_1}{\partial t}\right) + \frac{\partial}{\partial x}\left(\frac{\partial z_1}{\partial t}\right)$$

$$\Rightarrow \gamma_{zx} = \frac{\partial u}{\partial z} + \frac{\partial w}{\partial x} = \gamma_{xz}$$

The strain tensor can be express as

$$\epsilon_{ij} = \begin{bmatrix} \epsilon_{xx} & \frac{1}{2}\gamma_{xy} & \frac{1}{2}\gamma_{xz} \\ \frac{1}{2}\gamma_{yx} & \epsilon_{yy} & \frac{1}{2}\gamma_{yz} \\ \frac{1}{2}\gamma_{zx} & \frac{1}{2}\gamma_{zy} & \epsilon_{zz} \end{bmatrix}$$

$$\Rightarrow \epsilon_{ij} = \begin{bmatrix} \frac{\partial u}{\partial x} & \frac{1}{2}\left(\frac{\partial v}{\partial x}+\frac{\partial u}{\partial y}\right) & \frac{1}{2}\left(\frac{\partial u}{\partial z}+\frac{\partial w}{\partial x}\right) \\ \frac{1}{2}\left(\frac{\partial v}{\partial x}+\frac{\partial u}{\partial y}\right) & \frac{\partial v}{\partial y} & \frac{1}{2}\left(\frac{\partial w}{\partial y}+\frac{\partial v}{\partial z}\right) \\ \frac{1}{2}\left(\frac{\partial u}{\partial z}+\frac{\partial w}{\partial x}\right) & \frac{1}{2}\left(\frac{\partial w}{\partial y}+\frac{\partial v}{\partial z}\right) & \frac{\partial w}{\partial z} \end{bmatrix}$$

TRANSFORMATION OF STRAIN TENSORS

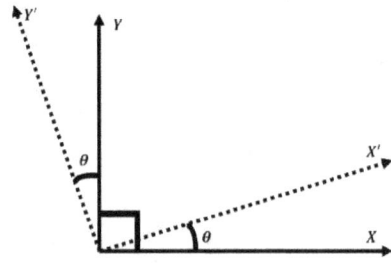

Let the two-dimensional strain components $\epsilon_{xx}, \epsilon_{yy}$, and γ_{xy} at O w.r.t. coordinate axes OX and OY. Let OX', OY' be

INTRODUCTION TO FLUID KINEMATICS

another set of orthogonal axes as shown in figure. Then $\epsilon_{x'x'}, \epsilon_{y'y'}$, and $\gamma_{x'y'}$ be the strain components w.r.t. new coordinates axes OX' and OY'. Then we have

$$x = l_1 x' + l_2 y' \quad \text{and} \quad y = m_1 x' + m_2 y'$$

Then velocity components u', v' w.r.t. new coordinates axes OX' and OY' are given by

$$u' = l_1 u + m_1 v \quad \text{and} \quad v' = l_2 u + m_2 v$$

The transformed rate of strain components $\epsilon_{x'x'}, \epsilon_{y'y'}$ and $\gamma_{x'y'}$ in x', y' directions on the elementary area normal to OX', OY' are given by

$$\epsilon_{x'x'} = \frac{\partial u'}{\partial x'} = \frac{\partial u'}{\partial x}\frac{\partial x}{\partial x'} + \frac{\partial u'}{\partial y}\frac{\partial y}{\partial x'}$$

$$\Rightarrow \epsilon_{x'x'} = \frac{\partial u'}{\partial x'} = \left(l_1 \frac{\partial u}{\partial x} + m_1 \frac{\partial v}{\partial x}\right) l_1 + \left(l_1 \frac{\partial u}{\partial y} + m_1 \frac{\partial v}{\partial y}\right) m_1$$

$$\Rightarrow \epsilon_{x'x'} = \frac{\partial u'}{\partial x'} = l_1^2 \frac{\partial u}{\partial x} + m_1^2 \frac{\partial v}{\partial y} + l_1 m_1 \left(\frac{\partial v}{\partial x} + \frac{\partial u}{\partial y}\right)$$

$$\Rightarrow \epsilon_{x'x'} = \frac{\partial u'}{\partial x'} = l_1^2 \epsilon_{xx} + m_1^2 \epsilon_{yy} + l_1 m_1 \gamma_{xy} \quad \text{--------(1)}$$

Similarly,

$$\epsilon_{y'y'} = \frac{\partial v'}{\partial y'} = \frac{\partial v'}{\partial x}\frac{\partial x}{\partial y'} + \frac{\partial v'}{\partial y}\frac{\partial y}{\partial y'}$$

$$\Rightarrow \epsilon_{yy} = \frac{\partial v'}{\partial y'} = \left(l_2 \frac{\partial u}{\partial x} + m_2 \frac{\partial v}{\partial x}\right) l_2 + \left(l_2 \frac{\partial u}{\partial y} + m_2 \frac{\partial v}{\partial y}\right) m_2$$

$$\Rightarrow \epsilon_{yy} = \frac{\partial v'}{\partial y'} = l_2^2 \frac{\partial u}{\partial x} + m_2^2 \frac{\partial v}{\partial y} + l_2 m_2 \left(\frac{\partial v}{\partial x} + \frac{\partial u}{\partial y}\right)$$

$$\Rightarrow \epsilon_{yy} = \frac{\partial v'}{\partial y'} = l_2^2 \epsilon_{xx} + m_2^2 \epsilon_{yy} + l_2 m_2 \gamma_{xy} \quad \text{--------(2)}$$

INTRODUCTION TO FLUID KINEMATICS

$$\gamma_{x'y'} = \frac{\partial v'}{\partial x'} + \frac{\partial u'}{\partial y'} = \frac{\partial v'}{\partial x}\frac{\partial x}{\partial x'} + \frac{\partial v'}{\partial y}\frac{\partial y}{\partial x'} + \frac{\partial u'}{\partial x}\frac{\partial x}{\partial y'} + \frac{\partial u'}{\partial y}\frac{\partial y}{\partial y'}$$

$$\Rightarrow \gamma_{x'y'} = \frac{\partial v'}{\partial x'} + \frac{\partial u'}{\partial y'} = \left(l_2 \frac{\partial u}{\partial x} + m_2 \frac{\partial v}{\partial x}\right)l_1 + \left(l_2 \frac{\partial u}{\partial y} + m_2 \frac{\partial v}{\partial y}\right)m_1$$

$$+ \left(l_1 \frac{\partial u}{\partial x} + m_1 \frac{\partial v}{\partial x}\right)l_2 + \left(l_1 \frac{\partial u}{\partial y} + m_1 \frac{\partial v}{\partial y}\right)m_2$$

$$\Rightarrow \gamma_{x'y'} = \frac{\partial v'}{\partial x'} + \frac{\partial u'}{\partial y'} = 2l_1 l_2 \frac{\partial u}{\partial x} + 2m_1 m_2 \frac{\partial v}{\partial y} +$$

$$l_1 m_2 \left(\frac{\partial u}{\partial y} + \frac{\partial v}{\partial x}\right) + l_2 m_1 \left(\frac{\partial u}{\partial y} + \frac{\partial v}{\partial x}\right)$$

$$\Rightarrow \gamma_{x'y'} = \frac{\partial v'}{\partial x'} + \frac{\partial u'}{\partial y'} = 2l_1 l_2 \in_{xx} + 2m_1 m_2 \in_{yy} +$$

$$(l_1 m_2 + l_2 m_1) \gamma_{xy} \quad \text{---------(3)}$$

The direction cosines of axes w.r.t. new coordinate axes are given as

	OX	OY
OX'	$l_1 = \cos\theta$	$m_1 = \cos\left(\frac{\pi}{2} - \theta\right)$ $\Rightarrow m_1 = \sin\theta$
OY'	$l_2 = \cos\left(\frac{\pi}{2} + \theta\right)$ $\Rightarrow l_2 = -\sin\theta$	$m_2 = \cos\theta$

Using the above mentioned table, equation (1), (2), and (3) can be re-written in terms of angle θ are follows:

INTRODUCTION TO FLUID KINEMATICS

$$\epsilon_{x'x'} = \cos^2\theta\, \epsilon_{xx} + \sin^2\theta\, \epsilon_{yy} + 2\sin\theta\cos\theta\, \gamma_{xy}$$

$$\epsilon_{y'y'} = \sin^2\theta\, \epsilon_{xx} + \cos^2\theta\, \epsilon_{yy} - 2\sin\theta\cos\theta\, \gamma_{xy}$$

$$\gamma_{x'y'} = -2\sin\theta\cos\theta\, \epsilon_{xx} + 2\sin\theta\cos\theta\, \epsilon_{yy} + \left(\cos^2\theta - \sin^2\theta\right)\gamma_{xy}$$

Recall high school knowledge, we have

$$\sin 2\theta = 2\sin\theta\cos\theta$$

$$\sin^2\theta = \frac{1}{2}(1 - \cos 2\theta)$$

$$\cos^2\theta = \frac{1}{2}(1 + \cos 2\theta)$$

Using above relations, transformed strain components can be re-written as

$$\epsilon_{x'x'} = \frac{1}{2}(\epsilon_{xx} + \epsilon_{yy}) + \frac{1}{2}(\epsilon_{xx} - \epsilon_{yy})\cos 2\theta + \gamma_{xy}\sin 2\theta \quad \text{------(4)}$$

$$\epsilon_{y'y'} = \frac{1}{2}(\epsilon_{xx} + \epsilon_{yy}) - \frac{1}{2}(\epsilon_{xx} - \epsilon_{yy})\cos 2\theta - \gamma_{xy}\sin 2\theta \quad \text{------(5)}$$

$$\gamma_{x'y'} = -\frac{1}{2}(\epsilon_{xx} - \epsilon_{yy})\sin 2\theta + \gamma_{xy}\cos 2\theta \quad \text{---------------(6)}$$

It follows that

$$\epsilon_{x'x'} + \epsilon_{y'y'} = \epsilon_{xx} + \epsilon_{yy}$$

$$\epsilon_{x'x'}\epsilon_{y'y'} - \frac{1}{4}\gamma_{x'y'}^2 = \epsilon_{xx}\epsilon_{yy} - \frac{1}{4}\gamma_{xy}^2$$

The above expressions shows that rate of strain components are remain invariant for the transformation of coordinate axes, which consider as **invariants of rate of strain in two-dimensional coordinate system.**

RELATION BETWEEN STRESS AND STRAIN TENSORS

We know that stresses are created by either the translation or rotation, so we can say that stress tensor is depends on the variation in rate of strain tensor. In order to derive the expression for the relation between the stress and strain

INTRODUCTION TO FLUID KINEMATICS

tensor, following assumption are necessary to taken into consideration:

1. The stress components can be written as a linear function of the rate of strain components.
2. The relation between stress-components and rate of strain components are invariant to a coordinate transformation created either by rotation or by mirror reflection of axes.
3. The stress-components reduce to the hydrostatic pressure P when all the velocity gradients are zero.

Recall from the above discussed topic of transformation of stress tensor components, we have

$$\sigma_{x'x'} = \frac{1}{2}(\sigma_{xx} + \sigma_{yy}) + \frac{1}{2}(\sigma_{xx} - \sigma_{yy})\cos 2\theta + \sigma_{xy}\sin 2\theta \quad \text{------(1)}$$

$$\sigma_{y'y'} = \frac{1}{2}(\sigma_{xx} + \sigma_{yy}) - \frac{1}{2}(\sigma_{xx} - \sigma_{yy})\cos 2\theta - \sigma_{xy}\sin 2\theta \quad \text{------(2)}$$

$$\sigma_{x'y'} = -\frac{1}{2}(\sigma_{xx} - \sigma_{yy})\sin 2\theta + \sigma_{xy}\cos 2\theta \quad \text{---------------(3)}$$

Using first assumption, we have

$$\sigma_{xx} = P_1\varepsilon_{xx} + Q_1\varepsilon_{yy} + R_1\gamma_{xy} + S_1 \quad \text{--------(4)}$$
$$\sigma_{yy} = P_2\varepsilon_{xx} + Q_2\varepsilon_{yy} + R_2\gamma_{xy} + S_2 \quad \text{--------(5)}$$
$$\sigma_{xy} = P_3\varepsilon_{xx} + Q_3\varepsilon_{yy} + R_3\gamma_{xy} + S_3 \quad \text{--------(6)}$$

Where P_i, Q_i, R_i and S_i are arbitrary constants to be determined.

In the same way, transformed stress components $\sigma_{x'x'}, \sigma_{y'y'}$ and $\sigma_{x'y'}$ w.r.t. new axes OX', OY' are given by

$$\sigma_{x'x'} = P_1\varepsilon_{x'x'} + Q_1\varepsilon_{y'y'} + R_1\gamma_{x'y'} + S_1 \quad \text{------(7)}$$
$$\sigma_{y'y'} = P_2\varepsilon_{x'x'} + Q_2\varepsilon_{y'y'} + R_2\gamma_{x'y'} + S_2 \quad \text{-------(8)}$$
$$\sigma_{x'y'} = P_3\varepsilon_{x'x'} + Q_3\varepsilon_{y'y'} + R_3\gamma_{x'y'} + S_3 \quad \text{-------(9)}$$

INTRODUCTION TO FLUID KINEMATICS

The stress component $\sigma_{x'x'}$ in x', y' direction is given by equation (1) re-written as:

$$\sigma_{x'x'} = \frac{1}{2}(1+\cos 2\theta)\sigma_{xx} + \frac{1}{2}(1-\cos 2\theta)\sigma_{yy} + \sigma_{xy}\sin 2\theta \quad ----(10)$$

Using (4), (5), and (6), the equation (7) reduced to

$$\sigma_{x'x'} = \left[\frac{P_1}{2}(1+\cos 2\theta) + \frac{P_2}{2}(1-\cos 2\theta) + P_3 \sin 2\theta\right]\varepsilon_{xx}$$

$$+ \left[\frac{Q_1}{2}(1+\cos 2\theta) + \frac{Q_2}{2}(1-\cos 2\theta) + Q_3 \sin 2\theta\right]\varepsilon_{yy}$$

$$+ \left[\frac{R_1}{2}(1+\cos 2\theta) + \frac{R_2}{2}(1-\cos 2\theta) + R_3 \sin 2\theta\right]\gamma_{xy}$$

$$+ \left[\frac{S_1}{2}(1+\cos 2\theta) + \frac{S_2}{2}(1-\cos 2\theta) + S_3 \sin 2\theta\right]$$

Recall from the above discussed topic of transformation of strain tensor components, we have

$$\varepsilon_{x'x'} = \frac{1}{2}(\varepsilon_{xx}+\varepsilon_{yy}) + \frac{1}{2}(\varepsilon_{xx}-\varepsilon_{yy})\cos 2\theta + \frac{1}{2}\gamma_{xy}\sin 2\theta \quad ----(11)$$

$$\varepsilon_{y'y'} = \frac{1}{2}(\varepsilon_{xx}+\varepsilon_{yy}) - \frac{1}{2}(\varepsilon_{xx}-\varepsilon_{yy})\cos 2\theta - \frac{1}{2}\gamma_{xy}\sin 2\theta \quad ----(12)$$

$$\gamma_{y'y'} = (\varepsilon_{yy}-\varepsilon_{xx})\sin 2\theta + \gamma_{xy}\cos 2\theta \quad --------(13)$$

Using (11), (12), and (13), the equation (7) reduced to

$$\sigma_{x'x'} = \left[\frac{P_1}{2}(1+\cos 2\theta) + \frac{Q_1}{2}(1-\cos 2\theta) - R_1 \sin 2\theta\right]\varepsilon_{xx}$$

$$+ \left[\frac{P_1}{2}(1-\cos 2\theta) + \frac{Q_1}{2}(1+\cos 2\theta) + R_1 \sin 2\theta\right]\varepsilon_{yy}$$

$$+ \left[\frac{P_1}{2}\sin 2\theta - \frac{Q_1}{2}\sin 2\theta + R_1 \cos 2\theta\right]\gamma_{xy} + S_1$$

On equating both expression for $\sigma_{x'x'}$, it follows that the

147

INTRODUCTION TO FLUID KINEMATICS

coefficients of $\varepsilon_{xx}, \varepsilon_{yy}, \gamma_{xy}$ must be same for all values of θ. Thus, we have

$$\frac{P_1}{2}(1+\cos 2\theta) + \frac{P_2}{2}(1-\cos 2\theta) + P_3 \sin 2\theta = \frac{P_1}{2}(1+\cos 2\theta) + \frac{Q_1}{2}(1-\cos 2\theta) - R_1 \sin 2\theta \quad \text{-----(14)}$$

$$\frac{Q_1}{2}(1+\cos 2\theta) + \frac{Q_2}{2}(1-\cos 2\theta) + Q_3 \sin 2\theta = \frac{P_1}{2}(1-\cos 2\theta) + \frac{Q_1}{2}(1+\cos 2\theta) + R_1 \sin 2\theta \quad \text{-------(15)}$$

$$\frac{R_1}{2}(1+\cos 2\theta) + \frac{R_2}{2}(1-\cos 2\theta) + R_3 \sin 2\theta = \frac{P_1}{2}\sin 2\theta - \frac{Q_1}{2}\sin 2\theta - R_1 \cos 2\theta \quad \text{--------(16)}$$

$$\frac{S_1}{2}(1+\cos 2\theta) + \frac{S_2}{2}(1-\cos 2\theta) + S_3 \sin 2\theta = S_1 \quad \text{---------(17)}$$

Solving above equations, we obtain as

$P_1 = Q_2 = P(say)$

$P_2 = Q_1 = Q(say)$

$P_3 = Q_3 = -R_1 = R_2 = R(say), \qquad R_3 = \frac{1}{2}(P_1 - Q_1)$

$S_1 = S_2 = S(say), \qquad S_3 = 0$

Using above relations, equation (4), (5), and (6) are reduced to

$$\sigma_{xx} = P\varepsilon_{xx} + Q\varepsilon_{yy} + R\gamma_{xy} + S \quad \text{-------(18)}$$

$$\sigma_{yy} = Q\varepsilon_{xx} + P\varepsilon_{yy} - R\gamma_{xy} + S \quad \text{-------(19)}$$

$$\sigma_{xy} = R(\varepsilon_{yy} - \varepsilon_{xx}) + \frac{1}{2}(P-Q)\gamma_{xy} \quad \text{------(20)}$$

Assume that $R = 0, S = -P$, then equations (18), (19), and (20) reduced to

$$\left.\begin{array}{l}\sigma_{xx} = P\varepsilon_{xx} + Q\varepsilon_{yy} - P \\ \sigma_{yy} = Q\varepsilon_{xx} + P\varepsilon_{yy} - P \\ \sigma_{xy} = \frac{1}{2}(P-Q)\gamma_{xy}\end{array}\right\} \quad \text{-------(21)}$$

Let us choose, a constant $\kappa = \frac{1}{2}(P-Q), \Rightarrow P = 2\kappa + Q$.

Thus, equation (21) reduced to

INTRODUCTION TO FLUID KINEMATICS

$$\left.\begin{array}{l}\sigma_{xx} = 2\kappa\varepsilon_{xx} + Q(\varepsilon_{xx} + \varepsilon_{yy}) - P \\ \sigma_{yy} = 2\kappa\varepsilon_{yy} + Q(\varepsilon_{xx} + \varepsilon_{yy}) - P \\ \sigma_{xy} = \kappa\gamma_{xy} = 2\kappa\varepsilon_{xy}\end{array}\right\} \quad \text{------(22)}$$

Where $Q = -\dfrac{2\kappa}{3}$ derived from the *Stokes's Law*.

Thus, viscous stress tensor for an incompressible Newtonian fluid is linearly proportional to the rate of strain tensor.

NAVIER-STOKES EQUATION

From the Newton's second law, we have

$$\sum F = M.a$$

Where F represented all different forces acts on an incompressible fluid and M is the total mass flow rate within a particular flow field which moves with the acceleration, a.

The above governing expression can be re-written as

$$\sum F = \frac{D}{Dt}(Mv) \quad (\because M = \text{constant})$$

$$\Rightarrow \sum F = \frac{\partial}{\partial t}(Mv) + \frac{\partial}{\partial x}(Mv)\frac{\partial x}{\partial t} + \frac{\partial}{\partial y}(Mv)\frac{\partial y}{\partial t} + \frac{\partial}{\partial z}(Mv)\frac{\partial z}{\partial t} \quad \text{------(1)}$$

RHS of equation (1) can be express in terms of convective derivative for an infinitesimal control volume whose volume is $\delta x \delta y \delta z$ flows with uniform density ρ, so we obtain as

$$(1) \Rightarrow \sum F = \rho\left[\frac{\partial v}{\partial t} + \left(\frac{\partial v}{\partial x}.\frac{\partial x}{\partial t}\right)\hat{i} + \left(\frac{\partial v}{\partial y}.\frac{\partial y}{\partial t}\right)\hat{j} + \left(\frac{\partial v}{\partial z}.\frac{\partial z}{\partial t}\right)\hat{k}\right]\delta x \delta y \delta z$$

$$\Rightarrow \sum F = \rho\left[\frac{\partial v}{\partial t} + \left(\frac{\partial v}{\partial x}.v_x\right)\hat{i} + \left(\frac{\partial v}{\partial y}.v_y\right)\hat{j} + \left(\frac{\partial v}{\partial z}.v_z\right)\hat{k}\right]\delta x \delta y \delta z \quad \text{------(2)}$$

Moreover, there are three driving forces acts on the fluid flow namely, gravitational force, pressure force, and viscous force. Thus, LHS of (2) can be express as

INTRODUCTION TO FLUID KINEMATICS

$$F_{grv}+F_{pressure}+F_{visc}=\rho\left[\frac{\partial v}{\partial t}+\left(v_x\frac{\partial v}{\partial t}\right)\hat{i}+\left(v_y\frac{\partial v}{\partial t}\right)\hat{j}+\left(v_z\frac{\partial v}{\partial t}\right)\hat{k}\right]\delta x\delta y\delta z \quad —(3)$$

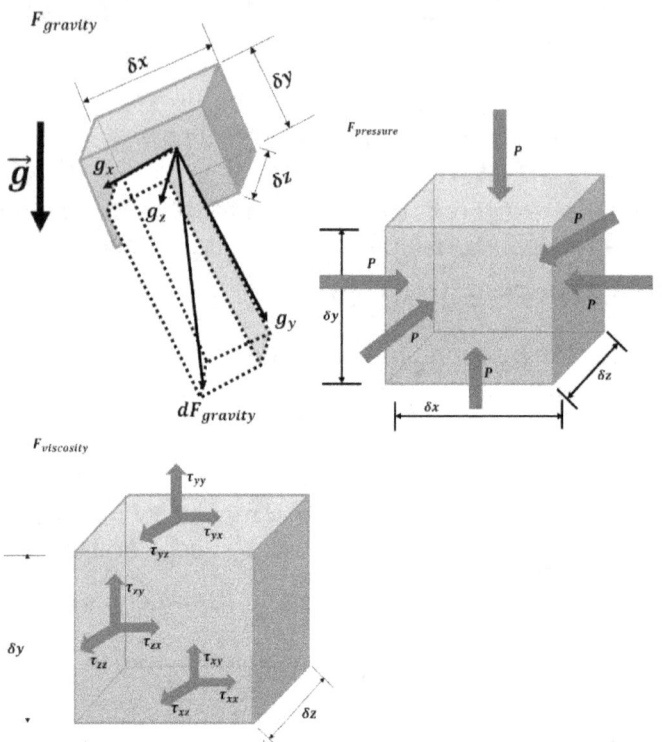

Gravitational force is the self-induced force which acts on the centre of mass of an infinitesimal control volume and it is given by

$$F_{gravity} = \text{mass} \times \text{acceleration} = (\rho\delta x\delta y\delta z)g$$
$$\Rightarrow F_{grv} = \rho g \delta x \delta y \delta z$$

Pressure force is the normal stress which acts inward to the surface of an infinitesimal control volume and it is given by

$$F_{pressure} = -\nabla P.\delta V$$
$$\Rightarrow F_{pressure} = -\nabla P.\delta x \delta y \delta z$$

Viscosity of the fluid induced a viscous forces which acts on the surface of an infinitesimal control volume in outward

INTRODUCTION TO FLUID KINEMATICS

direction. The viscous force, shear stress provides some amount of frictional force to the flow of fluid and it is given by

$$F_{vis\cos ity} = \nabla \tau.\delta V = \nabla \tau \delta x \delta y \delta z$$

Since Young's modulus is zero for the fluids, thus shear stresses induces shear strain rates in particular flow field. So, we have

$$\tau_{xy} = \tau_{yx} = \mu(\dot{\varepsilon}_{xy} + \dot{\varepsilon}_{yx}) = \mu\left(\frac{\partial}{\partial t}\frac{\partial y}{\partial x} + \frac{\partial}{\partial t}\frac{\partial x}{\partial y}\right)$$

$$\Rightarrow \tau_{xy} = \tau_{yx} = \mu\left(\frac{\partial}{\partial x}\frac{\partial y}{\partial t} + \frac{\partial}{\partial y}\frac{\partial x}{\partial t}\right) = \mu\left(\frac{\partial v_y}{\partial x} + \frac{\partial v_x}{\partial y}\right)$$

$$\tau_{yz} = \tau_{zy} = \mu(\dot{\varepsilon}_{yz} + \dot{\varepsilon}_{zy}) = \mu\left(\frac{\partial}{\partial t}\frac{\partial z}{\partial y} + \frac{\partial}{\partial t}\frac{\partial y}{\partial z}\right)$$

$$\Rightarrow \tau_{yz} = \tau_{zy} = \mu\left(\frac{\partial}{\partial y}\frac{\partial z}{\partial t} + \frac{\partial}{\partial z}\frac{\partial z}{\partial t}\right) = \mu\left(\frac{\partial v_z}{\partial y} + \frac{\partial v_y}{\partial z}\right)$$

$$\tau_{xz} = \tau_{zx} = \mu(\dot{\varepsilon}_{xz} + \dot{\varepsilon}_{zx}) = \mu\left(\frac{\partial}{\partial t}\frac{\partial z}{\partial x} + \frac{\partial}{\partial t}\frac{\partial x}{\partial z}\right)$$

$$\Rightarrow \tau_{xz} = \tau_{zx} = \mu\left(\frac{\partial}{\partial x}\frac{\partial z}{\partial t} + \frac{\partial}{\partial z}\frac{\partial x}{\partial t}\right) = \mu\left(\frac{\partial v_z}{\partial x} + \frac{\partial v_x}{\partial z}\right)$$

$$\tau_{xx} = -\frac{2}{3}\mu\nabla.v + 2\mu\frac{\partial v_x}{\partial x}$$

$$\tau_{yy} = -\frac{2}{3}\mu\nabla.v + 2\mu\frac{\partial v_y}{\partial y}$$

$$\tau_{zz} = -\frac{2}{3}\mu\nabla.v + 2\mu\frac{\partial v_z}{\partial z}$$

Then equation (3) can be express as

INTRODUCTION TO FLUID KINEMATICS

$$[\rho g - \nabla P + \nabla \tau]\delta x \delta y \delta z$$

$$= \rho \left[\frac{\partial v}{\partial t} + \left(v_x \frac{\partial v}{\partial t}\right)\hat{i} + \left(v_y \frac{\partial v}{\partial t}\right)\hat{j} + \left(v_z \frac{\partial v}{\partial t}\right)\hat{k} \right] \delta x \delta y \delta z$$

$$\Rightarrow \rho g - \nabla P + \nabla \tau = \rho \left[\frac{\partial v}{\partial t} + \left(v_x \frac{\partial v}{\partial t}\right)\hat{i} + \left(v_y \frac{\partial v}{\partial t}\right)\hat{j} + \left(v_z \frac{\partial v}{\partial t}\right)\hat{k} \right]$$

Rearrange the above relation into 3 components with reference to the three dimensional coordinate system as

$$\rho g_x - \frac{\partial P}{\partial x} + \frac{\partial \tau_{xx}}{\partial x} + \frac{\partial \tau_{yx}}{\partial y} + \frac{\partial \tau_{zx}}{\partial z} = \rho \left(\frac{\partial v_x}{\partial t} + v_x \frac{\partial v_x}{\partial x} + v_y \frac{\partial v_x}{\partial y} + v_z \frac{\partial v_x}{\partial z} \right)$$

$$\rho g_y - \frac{\partial P}{\partial y} + \frac{\partial \tau_{xy}}{\partial x} + \frac{\partial \tau_{yy}}{\partial y} + \frac{\partial \tau_{zy}}{\partial z} = \rho \left(\frac{\partial v_y}{\partial t} + v_x \frac{\partial v_y}{\partial x} + v_y \frac{\partial v_y}{\partial y} + v_z \frac{\partial v_y}{\partial z} \right)$$

$$\rho g_z - \frac{\partial P}{\partial z} + \frac{\partial \tau_{xz}}{\partial x} + \frac{\partial \tau_{yz}}{\partial y} + \frac{\partial \tau_{zz}}{\partial z} = \rho \left(\frac{\partial v_z}{\partial t} + v_x \frac{\partial v_z}{\partial x} + v_y \frac{\partial v_z}{\partial y} + v_z \frac{\partial v_z}{\partial z} \right)$$

$$\Rightarrow \rho g_x - \frac{\partial P}{\partial x} + \mu \left(\frac{\partial^2 v_x}{\partial x^2} + \frac{\partial^2 v_x}{\partial y^2} + \frac{\partial^2 v_x}{\partial z^2} \right)$$

$$= \rho \left(\frac{\partial v_x}{\partial t} + v_x \frac{\partial v_x}{\partial x} + v_y \frac{\partial v_x}{\partial y} + v_z \frac{\partial v_x}{\partial z} \right)$$

$$\Rightarrow \rho g_y - \frac{\partial P}{\partial y} + \mu \left(\frac{\partial^2 v_y}{\partial x^2} + \frac{\partial^2 v_y}{\partial y^2} + \frac{\partial^2 v_y}{\partial z^2} \right)$$

$$= \rho \left(\frac{\partial v_y}{\partial t} + v_x \frac{\partial v_y}{\partial x} + v_y \frac{\partial v_y}{\partial y} + v_z \frac{\partial v_y}{\partial z} \right)$$

$$\Rightarrow \rho g_z - \frac{\partial P}{\partial z} + \mu \left(\frac{\partial^2 v_z}{\partial x^2} + \frac{\partial^2 v_z}{\partial y^2} + \frac{\partial^2 v_z}{\partial z^2} \right)$$

$$= \rho \left(\frac{\partial v_z}{\partial t} + v_x \frac{\partial v_z}{\partial x} + v_y \frac{\partial v_z}{\partial y} + v_z \frac{\partial v_z}{\partial z} \right)$$

The set of non-linear PDE is known as **Navier-Stoke's Equations** which consider as cornerstone of fluid dynamic.

INTRODUCTION TO FLUID KINEMATICS

BOOSTER CAPSULE : TORQUE

We know that a system gain acceleration due to the application of an external force and it possess translational motion. Similarly, a system possess rotational motion when an external force acts on it from the axis of rotation of the system. The product of the external force and the distance of the point of application of the force from the axis of rotation is represented by a physical quantity called as **torque**. Thus, it is defined as

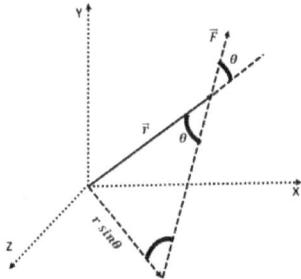

Torque acts on a system is the vector product of the magnitude of the applied force and the distance of the applied force from the axis of rotation of the system. It is also refer to Moment of forces.

Figure depicts a particle positioned at the point P w.r.t. the point O in two-dimension cartesian system, i.e., X-Y plane. Assume that $\vec{F}$ be the external force acts on the particle at an angle θ along the direction of the position vector $\vec{r}$. Thus, the torque acts on the particle is given by

$$\vec{\tau} = \vec{r} \times \vec{F}$$

The magnitude of the torque is given by $\tau = (r \sin \theta) F$ where distance $r \sin \theta$ is considered as lever arm as depicts in figure.

Direction of Torque

The torque produce is always perpendicular to the plane covered $\vec{r}$ and $\vec{F}$. The direction of the torque is determined by Right Hand Rule as explained below:

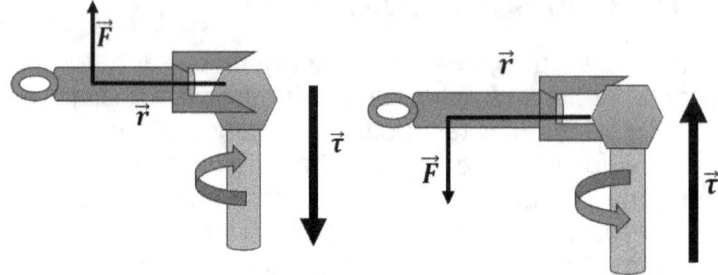

The torque produce in upward direction when applied force $\vec{F}$ acts at a distance $\vec{r}$ from the bolt in anticlockwise. Thus, bolt will be loosen as shown in figure. On other hand, same amount of torque produced in downward direction when applied force $\vec{F}$ acts at a distance $\vec{r}$ from the bolt in clockwise direction. As a result, the bolt will be tightened as shown in figure.

❖ Consider the case when $\theta = 0$, it means that the applied force acts in the direction of position vector and the amount of torque produced is $\tau = (r\sin 0°)F = 0.F = 0$. Visualize an external force $\vec{F}$ acts on the door to pull it away from the hinges as shown in figure. Here, no torque produce and thus the door will not rotate.

❖ Consider the case when $\theta = 90°$, it means that the applied force is perpendicular to the direction of position vector and the amount of torque produced is $\tau = (r\sin 90°)F = rF$ (maximum). Visualize an external force $\vec{F}$ applied at the end of the door perpendicular to i9ts plane as shown in figure. Here, maximum torque produce and thus door will rotate easily.

❖ Consider the case when $r = 0$, it means that the applied force acts on the axis of rotation and torque produce is $\tau = (0.\sin\theta)F = 0.F = 0$. Visualize an external force $\vec{F}$ acts on the door very close to

INTRODUCTION TO FLUID KINEMATICS

door's hinges as shown in figure. Here, no torque acts on the door and thus door will not rotate.

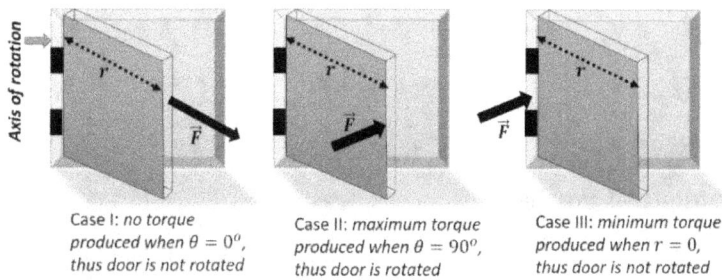

Case I: *no torque produced when* $\theta = 0°$, *thus door is not rotated*

Case II: *maximum torque produced when* $\theta = 90°$, *thus door is rotated*

Case III: *minimum torque produced when* $r = 0$, *thus door is not rotated*

Angular Momentum

Angular momentum of a particle is defined as the product of linear momentum of the particle and perpendicular distance of the particle from the axis of rotation. It is define as

$$\vec{w} = \vec{r} \times \vec{L}$$

where $\vec{L} = m\vec{v}$, linear momentum of the particle

$\vec{r}$ = position vector of the particle from the axis of rotation.

Differentiating $\vec{w}$ w.r.t. 't', we get

$$\frac{d\vec{w}}{dt} = \frac{d}{dt}(\vec{r} \times \vec{L}) = \frac{d\vec{r}}{dt} \times \vec{L} + \vec{r} \times \frac{d\vec{L}}{dt}$$

$$\Rightarrow \frac{d\vec{w}}{dt} = \vec{v} \times m\vec{v} + \vec{r} \times \vec{F} \qquad \left(\because \frac{d\vec{r}}{dt} = \vec{v} \text{ and } \frac{d\vec{L}}{dt} = \vec{F} \right)$$

$$\Rightarrow \frac{d\vec{w}}{dt} = m(\vec{v} \times \vec{v}) + \vec{r} \times \vec{F}$$

$$\Rightarrow \frac{d\vec{w}}{dt} = \vec{r} \times \vec{F} \qquad [\because \vec{v} \times \vec{v} = 0]$$

$$\Rightarrow \frac{d\vec{w}}{dt} = \vec{\tau} \qquad [\because \vec{r} \times \vec{F} = \vec{\tau}]$$

Hence, torque is equivalent to the rate of change of angular momentum of the particle.

Correlation between angular momentum and torque

INTRODUCTION TO FLUID KINEMATICS

It states that whenever the external torque acts on the body is zero then angular momentum of a body remains constant. Mathematically, it can be express as

$$\frac{d\vec{w}}{dt} = 0 \quad [\because \tau = 0]$$

$$\Rightarrow \vec{w} = \text{constant}$$

Moment of Inertia

As we know from the Newton's first law of motion that the body is at rest or moves with uniform motion unless a force is applied on it. So, we can say that If mass of the body is heavier, then greater amount of force to be applied on the body to increase its linear acceleration. This property of the body which can change its state of linear motion is refer to inertia. Thus, in linear motion, mass of the body is a measure of inertia. In the same way, in rotational motion, a torque is required in order to produce angular acceleration of the body. So, we can say that if a body rotating about its axis unable to produce any change in its rotational motion by itself, then an external torque is applied on the body to increase its angular acceleration. This property of the body by virtue of which can change it state of rotational motion is refer to moment of inertia. It can be assumed as rotational inertia as *moment of inertia plays the same role in rotational motion as mass does in linear motion.*

Definition: Moment of inertia of a body about an axis of rotation is defined as the product of mass of the body and square of the distance from axis of rotation. It is denoted by I which is defined as

$$I = mr^2$$

Correlation between Moment of Inertia and Torque

The Torque acts on a body due to applied force $\vec{F}$ about axis of rotation at a distance $\vec{r}$ is given by $\vec{\tau} = \vec{r} \times \vec{F}$

$$\Rightarrow \tau = (r \sin\theta) F = rF \quad (\because \vec{F} \perp \vec{r})$$

$$\therefore \tau = rma \quad (\because F = ma)$$
$$\Rightarrow \tau = mr^2\alpha \quad (\because a = r\alpha)$$
$$\Rightarrow \tau = I\alpha \quad (\because I = mr^2)$$

Correlation between Moment of Inertia and Angular momentum

Let us assume that a body of mass m rotating with an angular velocity $\vec{\omega}$ along $z-$ axis whose position vector is $\vec{r}$ as shown in figure. The linear momentum $\vec{w}$ acts on the body is equal to

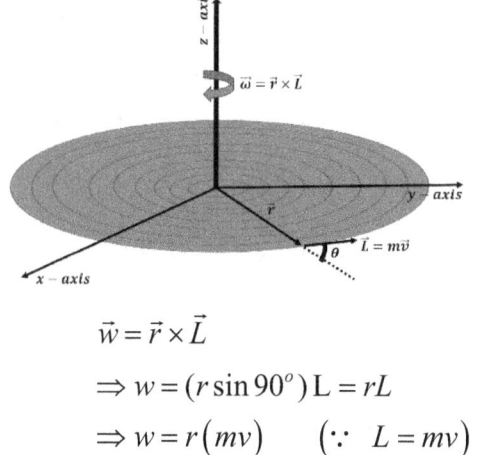

$$\vec{w} = \vec{r} \times \vec{L}$$
$$\Rightarrow w = (r\sin 90°)L = rL$$
$$\Rightarrow w = r(mv) \quad (\because L = mv)$$
$$\Rightarrow w = mr^2\omega \quad (\because v = r\omega)$$
$$\Rightarrow \vec{w} = I\vec{\omega} \quad (\because I = mr^2)$$

The directions of $\vec{w}$ and $\vec{\omega}$ are along the axis of rotation.

Law of Conservation of Angular Momentum

We know that angular momentum of a body is defined as
$$\vec{w} = I\vec{\omega}$$
Differentiating both sides w.r.t. t, we obtain as

$$\frac{d}{dt}(\vec{w}) = \frac{d}{dt}(I\vec{\omega})$$

$$\Rightarrow \frac{d}{dt}(\vec{w}) = \vec{\tau} = \frac{d}{dt}(I\vec{\omega})$$

If torque acts on the body is zero, then $\tau = 0 \Rightarrow \frac{d}{dt}(I\vec{\omega}) = 0$

On integrating above expression, we obtain as

$$I\vec{\omega} = \text{constant}$$
$$\Rightarrow \vec{w} = I\vec{\omega} = \text{constant}$$

Thus, whenever moment of inertia changes from I_1 to I_2 and I_2 to I_3 due to the orientation of mass of the body, then angular velocity of the body changes from ω_1 to ω_2 and ω_2 to ω_3. Since

$$\vec{w} = I\vec{\omega} = \text{constant}$$
$$\Rightarrow I_1\omega_1 = I_2\omega_2 = I_3\omega_3$$

which is known as Law of conservation of angular momentum.

Practical applications of Law of Conservation of angular momentum

1. Angular motion of a tornado is very high

We know from the principle of conservation of angular momentum that product of moment of inertia and angular velocity remains constant during the rotational movement of the body. So, it can be express as

$$I\vec{\omega} = \text{constant}$$

$$\Rightarrow \vec{\omega} \propto \frac{1}{I} \Rightarrow \vec{\omega} \uparrow \text{ when } I \downarrow \text{ or } \vec{\omega} \downarrow \text{ when } I \uparrow$$

The moment of inertia of a body is defined as

INTRODUCTION TO FLUID KINEMATICS

$I = mr^2$

$\Rightarrow I \propto r$ and $I \propto m \Rightarrow I \downarrow$ when $r \downarrow$ or $I \uparrow$ when $r \uparrow$

Thus, during tornado, rotational movement of air is very close to the central point as a result the moment of inertia of the air decreases which increases the angular velocity of the tornado.

2. **Diver curls his body when he jumps from the spring board to swimming pool**

When diver curls his body by rolling his arms and legs into different postures, this activity decreases moment of inertia of diver's body as a result angular speed increases according to the principle of conservation of angular momentum. Thus, diver jump into swimming pool with greater speed.

3. **Mercury revolves around the sun with high orbital speed**

Mercury revolves around the sun with the orbital speed of 47.87 km/sec (29.8 mi/sec), which takes only 88 days to complete one revolution, compared to 365 days taken by Earth planet. This is because as the planet comes close to the sun, its moment of inertia decreases and as a result planet's orbital velocity increases according to the principle of conservation of angular momentum. Hence, mercury, the closest planet to the sun which revolving around the sun in its elliptical orbit with highest speed.

4. **Dancer stretching their legs to speed up their dancing performance and vice-versa**

As dancer stretches her legs out, her moment of inertia increases and hence angular speed decreases according to the principle of conservation of angular momentum. In the same way, when she collapse her legs towards her body, her moment of inertia decreases and angular speed of dancer's body increases.

HIGHER ORDER THINKING SKILL QUESTION

Q. Calculate the velocity and pressure distribution of the flow through vacuum cleaner.

Solution: Consider the air flow into the floor attachment orifice of a typical household vacuum cleaner as shown in figure. The width of the orifice at inlet slot is $w = 1$ mm and its length is $L = 35$ cm. The inlet slot is held at a distance $d = 2$ cm above the floor as shown in figure. The total volume flow rate through the vacuum inlet pipe is $Q = 110$ Lt/sec. We are to consider the following assumptions in order to determine the flow field in the centre plane of vacuum cleaner attachment, and predict the velocity and pressure distribution along the x-axis.

1. The inviscid air flow is steady and incompressible.
2. The flow field plotted in XY-plane, so flow is refer to be a two-dimensional planar.
3. The flow field in the center plane of a vacuum cleaner is assumed to be irrotational.
4. The room is infinitely large and free of air currents such that they will not affect on flow field.

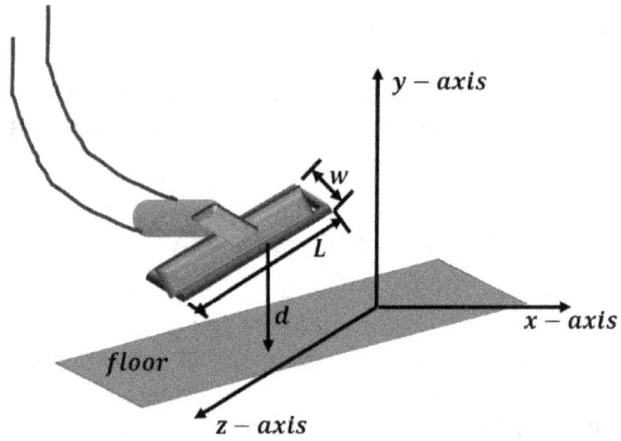

Consider flow through vacuum cleaner attachment as a sink (a source with negative strength), located at distance above the floor (x-axis). We plot the flow as the sink at a point in xy-plane with the coordinates. We also ignore any effects of the vacuum cleaner attachment. The strength of the sink can be calculated by dividing total volume flow rate by the length of the slot.

$$k = \frac{Q}{L} = \frac{-0.11}{0.35} = -0.314 \text{ m}^2/\text{sec}$$

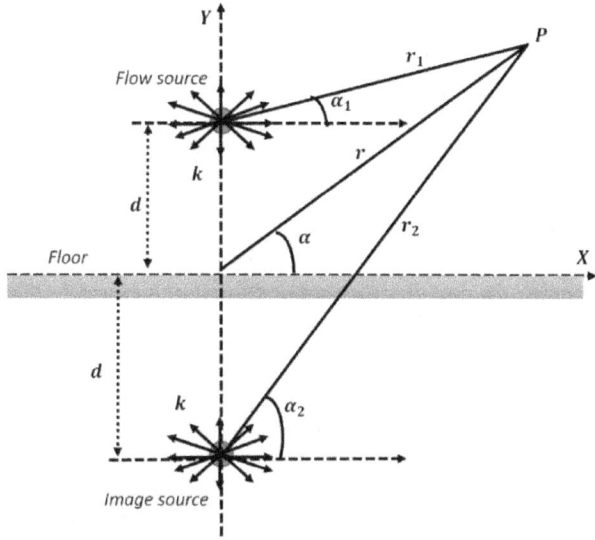

Since air flow into the sink from all direction, including up through the floor. For convenience, we introduce another elementary irrotational flow to the flow field such a way that it will be identical sink below the floor at point $(0,-d)$ and assumed this as image sink. Since x-axis is a line of symmetry, the x-axis is itself a streamline of the flow as the floor. The two sinks of strength of an irrotational flow are shown in figure. Figure depicts that top one is flow sink and shows the suction point of the vacuum cleaner attachment while bottom one is the image sink.

The streamlines for the sink at $(0,d)$ is given by

$$\psi_1 = \frac{k}{2\pi}\alpha_1 \text{ where } \alpha_1 = \tan^{-1}\left(\frac{y-d}{x}\right)$$

The streamlines for the image sink at $(0,-d)$ is given by

INTRODUCTION TO FLUID KINEMATICS

$$\psi_2 = \frac{k}{2\pi}\alpha_2 \text{ where } \alpha_2 = \tan^{-1}\left(\frac{y+d}{x}\right)$$

The composite stream function is given by

$$\psi = \psi_1 + \psi_2$$

$$\Rightarrow \psi = \frac{k}{2\pi}(\alpha_1 + \alpha_2) \Rightarrow \alpha_1 + \alpha_2 = \frac{2\pi\psi}{k}$$

$$\Rightarrow \tan(\alpha_1 + \alpha_2) = \tan\left(\frac{2\pi\psi}{k}\right)$$

$$\tan(\alpha_1 + \alpha_2) = \frac{\tan\alpha_1 + \tan\alpha_2}{1 - \tan\alpha_1 \tan\alpha_2} = \tan\left(\frac{2\pi\psi}{k}\right)$$

$$\Rightarrow \tan\left(\frac{2\pi\psi}{k}\right) = \frac{2xy}{x^2 - y^2 + d^2}$$

$$\Rightarrow \psi = \frac{k}{2\pi}\tan^{-1}\left(\frac{2xy}{x^2 - y^2 + d^2}\right)$$

Transform the above expression into cylindrical coordinates, we obtain as

$$\psi = \frac{k}{2\pi}\tan^{-1}\left(\frac{\sin 2\alpha}{\cos 2\alpha + \frac{1}{r_o^2}}\right) \text{ where } r_o = \frac{r}{d}$$

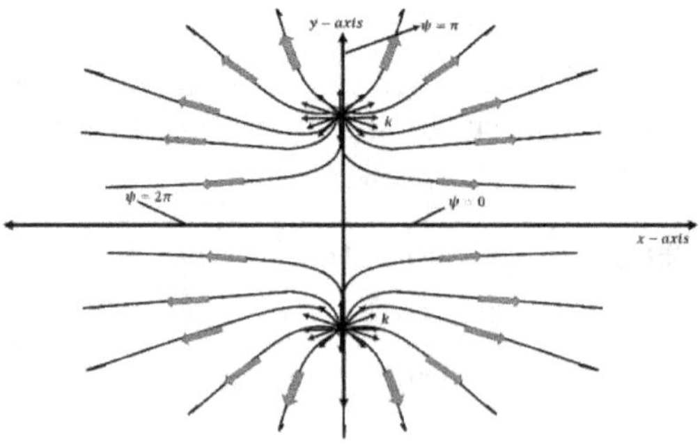

Curves showing as the function of are plotted in the figure. Clearly observed from the figure that x-axis is the line of symmetry such as the air flow produced by the upper sink is above the x-axis while air suction produced at the lower sink is below the x-axis. Thus, x-axis acts like as streamline separating two flow field. As we knew that difference in the value of streamline with another in any planar flow is equivalent to the volume flow rate per unit width flowing between the two streamlines. So, in this case of the vacuum cleaner, set of ψ equal to zero along positive x-axis whereas on the negative x-axis must be equivalent to the volume flow rate per unit width sucks through the upper sink which represents as

$$\psi_{-ve\ x-axis} + \underbrace{\psi_{+ve\ x-axis}}_{\equiv 0} = k$$

$$\Rightarrow \psi_{-ve\ x-axis} = 2\pi$$

To determine the velocity distribution on the floor, we use vector algebra and results illustrated in the figure below. The figure depicts the resultant velocity acts horizontal direction at any location on x-axis due to line of symmetry (x-axis). The velocity induced from the upper sink has magnitude $\dfrac{k}{2\pi r_1}$ and its direction is in line with r_1 as shown in figure.

On the same way, velocity induced from the image sink has same magnitude but its direction is in line with r_2. The vector sum of these two velocities gives the resultant velocity, $\vec{v}$ acts along x-axis and vertical components cancel out. Thus, the resultant velocity along x-axis is given by

$$v = \frac{kx}{\pi\left(x^2 + d^2\right)}$$

INTRODUCTION TO FLUID KINEMATICS

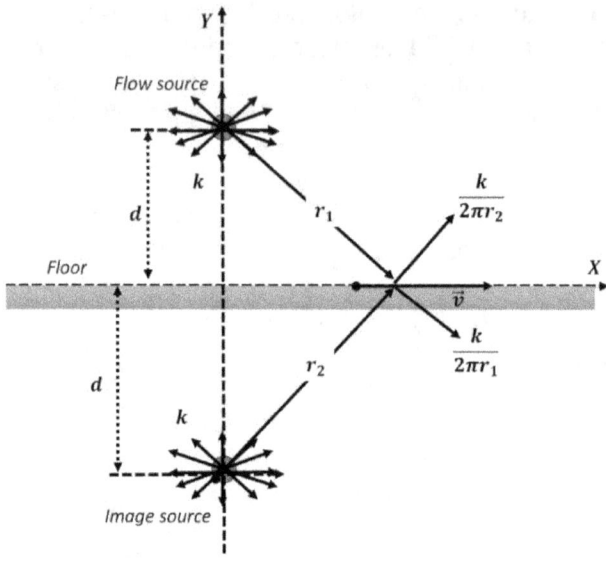

Using assumption 3, we apply Bernoulli's equation to the flow field, we get

$$\frac{P}{\rho} + \frac{1}{2}v^2 = \text{constant} = \frac{P_\infty}{\rho} + \frac{1}{2}\underbrace{v_\infty^2}_{\to 0}$$

$$\Rightarrow P - P_\infty = \frac{1}{2}\rho v^2$$

To calculate the pressure coefficient, we need to introduce a preferred velocity, $v_p = -\dfrac{k}{d}$. Then we define pressure coefficient, C_p as

$$C_p = \frac{P - P_\infty}{\frac{1}{2}\rho v_p^2} = -\frac{v^2}{v_p^2}$$

$$\Rightarrow C_p = -\frac{d^2 v^2}{k^2}$$

$$\Rightarrow C_p = -\frac{d^2 x^2}{\pi^2 \left(x^2 + d^2\right)^2}$$

INTRODUCTION TO FLUID KINEMATICS

Let us choose nondimensional variables for the axial velocity and distance as

$$\delta = \frac{v}{v_p} = -\frac{vd}{k} \quad \text{and} \quad y = \frac{x}{d}$$

Then expression for axial velocity, δ and pressure coefficient, C_p in dimensionless form are

$$\delta = -\frac{1}{\pi}\cdot\frac{y}{1+y^2} \qquad C_p = -\left(\frac{1}{\pi}\cdot\frac{y}{1+y^2}\right)^2 = -\delta^2$$

Curves showing δ and C_p as functions of y are plotted in the figure. It is clearly observed from the figure that increases from 0 at $y = -\infty$ to a maximum value of about 0.17 at $y = -1$. the velocity is positive for the negative value of y as air is being sucked into vacuum cleaner. Since the velocity increases, pressure decrease as a direct consequence of Bernoulli's equation. Thus, pressure coefficient C_p is 0 at $y = -\infty$ decreases to its minimum value of about -0.0253 at $y = -1$. Between $y = -1$ and $y = 0$ the velocity decreases to zero while pressure increases from the zero at the stagnation point which lies below the vacuum cleaner orifice. To the right of the orifice, the velocity, is antisymmetric, while the pressure is symmetric.

INTRODUCTION TO FLUID KINEMATICS

5. VORTICITY

INTRODUCTION

We often observe the rotation of fluid (air or liquids) in cases of cyclones, flows through turbomachinery (fans, turbines, compressors etc.), and wakes. All these cases formed flow field consist of rotational region as well as irrotational region. Physically, fluid elements in the rotational region of the flow field rotates and angular velocity vary with the change in viscosity, temperature gradients, and other non-uniform phenomena. However, fluid elements outside the rotational region of the flow are not moving with flow and remains in irrotational region unless any external non-uniform phenomena occurs. Mathematically, the rate of rotation of the fluid elements explained with vorticity vector, whose direction can be calculated by using the right hand rule for cross product.

Let $V = u\,i + v\,j + w\,k$ be the fluid velocity such that at $curl\,V \neq 0$. Then the vector

$$\Omega = curl\,V$$

is called the vorticity vector.

The curl of the velocity field of a fluid, which is generally termed *vorticity*, is usually represented by the symbol Ω, so that $\Omega = \nabla \times V$

Vorticity components (components of spin)

If **q** be the velocity vector of a fluid particle, then the vector

INTRODUCTION TO FLUID KINEMATICS

quantity, Ω ($= curl\ q$), is called the *vorticity vector* or simply the *vorticity* and is a measure of the angular velocity of an infinitesimal element.

Let $\Omega = \Omega_x\,i + \Omega_y\,j + \Omega_z\,k$ so that ($\Omega_x, \Omega_y, \Omega_z$) are the *vorticity components* or the *components of the spin*. Then, if
$$q = u\,i + v\,j + w\,k.$$

$$\Omega_x = \frac{\partial w}{\partial y} - \frac{\partial v}{\partial z} \qquad \Omega_y = \frac{\partial u}{\partial z} - \frac{\partial w}{\partial x} \qquad \Omega_z = \frac{\partial v}{\partial x} - \frac{\partial u}{\partial y}$$

If $\Omega_x, \Omega_y, \Omega_z$, are all zero, the motion is *irrotational* and the velocity function ϕ exists and if $\Omega_x, \Omega_y, \Omega_z$ are not all zero, the motion is *rotational*.

In case of two-dimensional motion, we know that $w = 0$ and u and v are functions of x and y only and hence for two-dimensional case,

$$\Omega_x = 0, \qquad \Omega_y = 0\ and \qquad \Omega_z = \frac{\partial v}{\partial x} - \frac{\partial u}{\partial y}$$

It follows that in two-dimensional motion there can be at the most only one component of spin and its axis is perpendicular to the plane of the motion.

Vortex line

A *vortex line* is a curve drawn in the fluid such that the tangent to it at every point is in the direction of the vorticity vector. The differential equations of the vortex lines are

$$\frac{dx}{\Omega_x} = \frac{dy}{\Omega_y} = \frac{dz}{\Omega_z}$$

In two-dimensional motion, since axis of rotation at every point is perpendicular to this plane of motion and hence the vortex lines must be all parallel (because all vortex line will be perpendicular to the plane of motion).

Vortex tube

The vortex lines passes through every points of a closed curve occupy a tubular surface in the fluid consider as **vortex tube**.

INTRODUCTION TO FLUID KINEMATICS

Vortex filament

A vortex tube consist of infinitesimal cross sectional surface area considered as **vortex filament.**

CONSTITUTIVE EQUATIONS

Let us assume a vortex filament whose axis of rotation is parallel to $z-axis$. There is no fluid motion in the $z-$direction such that $w = 0$ and u,v are independent of z, provided u,v and w are the components of fluid motion. Thus, we have

$$\frac{\partial u}{\partial z} = 0 \quad \text{and} \quad \frac{\partial v}{\partial z} = 0$$

Here vorticity defined as

$$\Omega_x = 0 \quad \Omega_y = 0 \quad \text{and} \quad \Omega_z = \frac{\partial v}{\partial x} - \frac{\partial u}{\partial y} \quad \text{-------(1)}$$

The equation of continuity is given by

$$\frac{\partial u}{\partial x} + \frac{\partial v}{\partial y} = 0 \Rightarrow \frac{\partial v}{\partial y} = -\frac{\partial u}{\partial x} \Rightarrow \frac{\partial v}{\partial y} = \frac{\partial}{\partial x}(-u) \quad \text{------(2)}$$

In this case, equation of flow defined as

$$\frac{dx}{u} = \frac{dy}{v} = \frac{dz}{0} \Rightarrow \frac{dx}{u} = \frac{dy}{v} \Rightarrow vdx - udy = 0 \quad \text{-----(3)}$$

The expression in (2) indicates that $vdx - udy$ must be perfect differential, $d\psi\,(say)$. Then we have

$$vdx - udy = d\psi$$

$$\Rightarrow vdx - udy = \frac{\partial \psi}{\partial x}dx + \frac{\partial \psi}{\partial y}dy$$

$$\Rightarrow v = \frac{\partial \psi}{\partial x} \quad \text{and} \quad u = -\frac{\partial \psi}{\partial y}$$

The equation (3) reduces to $vdx - udy = d\psi = 0 \Rightarrow \psi = $ constant.

Thus, ψ is the stream function of the fluid flow.

INTRODUCTION TO FLUID KINEMATICS

Then $\Omega_z = \dfrac{\partial v}{\partial x} - \dfrac{\partial u}{\partial y} = \dfrac{\partial}{\partial x}\left(\dfrac{\partial \psi}{\partial x}\right) - \dfrac{\partial}{\partial y}\left(-\dfrac{\partial \psi}{\partial y}\right) = \dfrac{\partial^2 \psi}{\partial x^2} + \dfrac{\partial^2 \psi}{\partial y^2}$

Hence, inside the vortex filament, vorticity is given by

$$\Omega_x = 0 \quad \Omega_y = 0 \quad \text{and} \quad \Omega_z = \dfrac{\partial^2 \psi}{\partial x^2} + \dfrac{\partial^2 \psi}{\partial y^2}$$

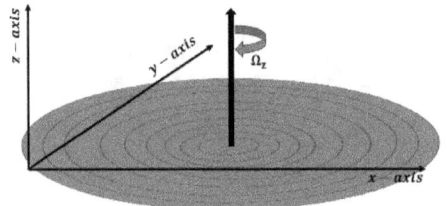

Moreover, $\Omega_z = 0$ outside the vortex filament, we can write this relation as

$$\dfrac{\partial^2 \psi}{\partial r^2} + \dfrac{1}{r}\dfrac{\partial \psi}{\partial r} + \dfrac{1}{r^2}\dfrac{\partial^2 \psi}{\partial \theta^2} = 0$$

Since vortex motion is symmetry about the origin, ψ is independent of θ and the above equation reduce to

$$\dfrac{\partial^2 \psi}{\partial r^2} + \dfrac{1}{r}\dfrac{\partial \psi}{\partial r} = 0$$

$$\Rightarrow \dfrac{1}{r}\dfrac{\partial}{\partial r}\left(r\dfrac{\partial \psi}{\partial r}\right) = 0$$

To obtain the expression for the stream function, above relation written in terms of total derivative and we obtain as

$$\dfrac{1}{r}\dfrac{d}{dr}\left(r\dfrac{d\psi}{dr}\right) = 0$$

Integrating above equation w.r.t. r, we get

$$r\frac{d\psi}{dr} = K \,(\text{constant})$$

$$\Rightarrow d\psi = K\frac{dr}{r}$$

$$\Rightarrow \int d\psi = K \int \frac{dr}{r}$$

$$\Rightarrow \psi = K\log|r| + K_1 \quad \text{------(4)}$$

Using initial condition when $r = 1$ then $\psi = 0$, the equation (4) gives $K_1 = 0$ and equ (4) reduced to

$$\psi = K\log|r|$$

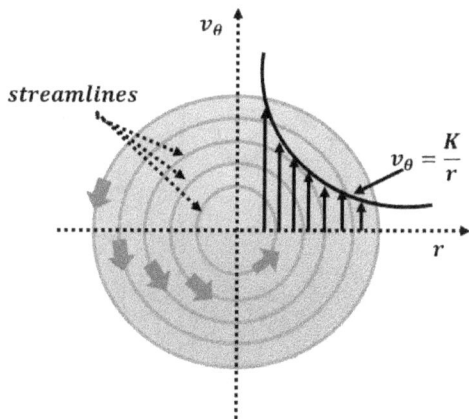

Assume that the fluid motion outside the vortex filament is irrotational, then velocity potential function ϕ exists. Thus, stream function in terms of velocity potential function given by

$$\frac{d\psi}{dr} = -\frac{1}{r}\frac{d\phi}{d\theta}$$

$$\Rightarrow \frac{d\psi}{dr} = \frac{K}{r} = -\frac{1}{r}\frac{d\phi}{d\theta} \quad \left[\because d\psi = K\frac{dr}{r}\right]$$

$$\Rightarrow K = -\frac{d\phi}{d\theta}$$

$$\Rightarrow d\phi = -Kd\theta$$

$$\Rightarrow \int d\phi = -K\int d\theta$$

$$\Rightarrow \phi = -K\theta \quad \text{----------(5)}$$

If $w = (\phi + i\psi)$ be the complex potential outside the vortex filament, so we have

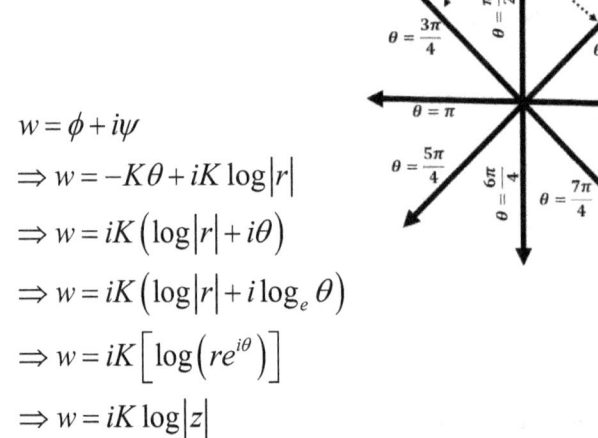

$w = \phi + i\psi$

$\Rightarrow w = -K\theta + iK\log|r|$

$\Rightarrow w = iK(\log|r| + i\theta)$

$\Rightarrow w = iK(\log|r| + i\log_e \theta)$

$\Rightarrow w = iK\left[\log(re^{i\theta})\right]$

$\Rightarrow w = iK\log|z|$

Let Γ be the circulation in closed curve embracing the vortex. Then, we have

$$\Gamma = \int_0^{2\pi} \left(-\frac{1}{r}\frac{d\phi}{d\theta}\right) r\,d\theta$$

$$\Rightarrow \Gamma = \int_0^{2\pi} \left(\frac{K}{r}\right) r\,d\theta$$

$$\Rightarrow \Gamma = K \int_0^{2\pi} d\theta = K[\theta]_0^{2\pi} = K(2\pi - 0)$$

$$\Rightarrow \Gamma = 2\pi K$$

$$\Rightarrow K = \frac{\Gamma}{2\pi}$$

Put the value of K in equation (5), we get

$$\phi = \frac{\Gamma}{2\pi}\theta$$

The above expression indicates that the velocity potential function, ϕ is a function of θ. For a particular value of θ, velocity potential function is constant. Thus, equipotential lines are shown in figure in radial direction.

VORTEX FLOW

The fluid flow where external force are applied to rotate the fluid within the system is known as vortex flow. The fluid motion of vortex flow is given by

$$v = \omega \times r$$

Where v = tangential component of fluid motion

ω = fluid motion in angular direction.

r = radius of fluid element from the axis of rotation.

Let us assume that an infinitesimal fluid element $PQRS$ rotating with the fluid motion v about an axis normal to the plane as shown in figure. Consider following parameter in order to derive the expression for the equation of motion of a vortex flow as

r = radial distance of the fluid element from the origin O.

INTRODUCTION TO FLUID KINEMATICS

δr = radial thickness created due to the rotation of the fluid element about an axis.

θ = angle makes by the fluid element.

a = cross-sectional area of the fluid element.

Mass of the fluid element = density × volume
$= \rho \times (a.\delta r) = \rho a \delta r$

Then centrifugal force acting on the fluid element
$= (\rho a \delta r) \dfrac{v^2}{2}$

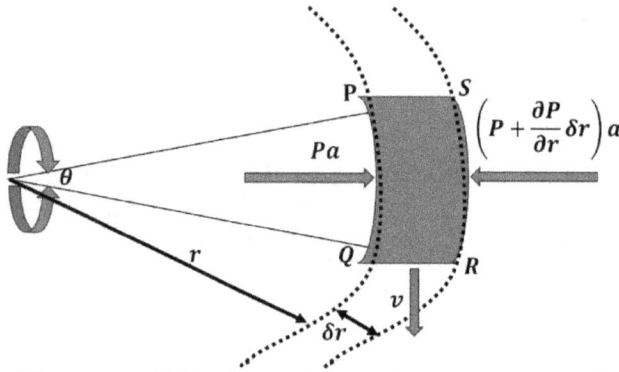

Moreover, fluid element experience pressure force from the both the sides, one is Pa from the side PQ and $\left(P + \dfrac{\partial P}{\partial r}\delta r\right) a$ from the side RS.

At the equilibrium state of the fluid element in field flow is given by

$\left(P + \dfrac{\partial P}{\partial r}\delta r\right) a - Pa = (\rho a \delta r)\dfrac{v^2}{2}$

$\Rightarrow \dfrac{\partial P}{\partial r}\delta r.a = (\rho a \delta r)\dfrac{v^2}{2}$

$\Rightarrow \dfrac{\partial P}{\partial r} = \rho \dfrac{v^2}{2}$

From the Hydrostatic Law, we have

$$\frac{\partial P}{\partial z} = -\rho g$$

Since pressure P is a function of r and z, so we can write

$$dP = \frac{\partial P}{\partial r} dr + \frac{\partial P}{\partial z} dz$$

Putting the value of $\frac{\partial P}{\partial r}$ and $\frac{\partial P}{\partial z}$ in the above expression, we get

$$dP = \rho \frac{v^2}{2} dr - \rho g dz$$

KELVIN CIRCULATION THEOREM

Statement: *The circulation in any closed enthalpy moving with the fluid is constant for all time provided the external forces are conservative and density is a function of pressure only.*

Proof: Let K be a closed enthalpy moving with the fluid so that K always comprise of same fluid particles. Let v be the motion of the fluid flow at any point A of the enthalpy and let r be its position vector. Then the circulation along the closed enthalpy K is given by

$$\Gamma = \int_K v.dr$$

$$\Rightarrow \frac{D}{Dt}(\Gamma) = \frac{D}{Dt}\left(\int_K v.dr\right) \quad \text{--------(1)}$$

Since the above integration is performed at constant time, reversing the order of integration and differentiation is justified. Then (1) can be written as

$$\frac{D}{Dt}\left(\int_K v.dr\right) = \int_K \frac{D}{Dt}(v.dr) \quad \text{--------(2)}$$

From the properties of derivative, we have

INTRODUCTION TO FLUID KINEMATICS

$$\frac{D}{Dt}(v.dr) = \left(\frac{D}{Dt}.v\right).dr + v.\left(\frac{D}{Dt}.dr\right)$$

$$\Rightarrow \frac{D}{Dt}(v.dr) = \left(\frac{D}{Dt}.v\right).dr + v.dv \quad \text{--------(3)}$$

The Euler's equation of motion is given by

$$\frac{D}{Dt}.v = E - \frac{1}{\rho}\nabla P$$

Since the external forces are conservative, so we can write $E = -\nabla F$

The Euler's equation of motion re-written as

$$\frac{D}{Dt}.v = -\nabla F - \frac{1}{\rho}\nabla P$$

$$\left(\frac{D}{Dt}.v\right)dr = -\nabla F.dr - \frac{1}{\rho}\nabla P.dr$$

$$\Rightarrow \left(\frac{D}{Dt}.v\right)dr = -dF - \frac{1}{\rho}.dP \quad \text{------(4)}$$

Also, we have, $v.dv = \frac{1}{2}d(v.v) = \frac{1}{2}dv^2 \quad \text{-------(5)}$

Using above relations given by (4) and (5), the equation(3) becomes

$$\frac{D}{Dt}(v.dr) = -dF - \frac{1}{\rho}dP + \frac{1}{2}dv^2$$

$$\Rightarrow \frac{D}{Dt}(v.dr) = d\left(\frac{1}{2}v^2 - F - \frac{P}{\rho}\right)$$

$$\Rightarrow \frac{D}{Dt}(\Gamma) = \frac{D}{Dt}\left(\int_K v.dr\right) = \int_K \frac{D}{Dt}(v.dr) = \int_K d\left(\frac{1}{2}v^2 - F - \frac{P}{\rho}\right)$$

$$\Rightarrow \frac{D}{Dt}(\Gamma) = \frac{1}{2}v^2 - F - \int_K \frac{dP}{\rho} \quad \text{-------(6)}$$

Where v, F and P are single-valued function of r, so R.H.S. of (6) vanishes. The equation (6) indicates that rate of change of flow along any closed enthalpy moving with the fluid. As a results the circulation in any closed enthalpy moving with the fluid is constant for all time.

IRROTATIONAL FLOW

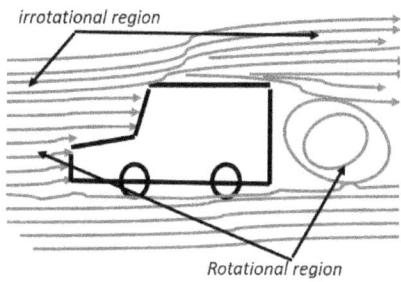

irrotational region

Rotational region

Figure depicts a situation in which the flow field comprise of two different regions , namely, an inviscid, rotational region and an inviscid, irrotational region. You must keep in mind that the assumption of irrotationality is considered for some regions of flow, but not for whole region of flow as shown in figure. Irrotational regions of flow are those region where there is no rotation or vortex line formation. Mostly, inviscid fluid which flows in open ended surface are considered as irrotational, and inviscid fluid flows in confined boundary area induce vortex lines and rotational region sustain.

When external forces are conservative and derivable from a single valued function and density is a function of pressure of the flow only, then the motion of inviscid fluid becomes irrotational.

From Stokes' theorem, the vorticity is given by

$$\Gamma = \int_C v.dr = \int_S \nabla \times v.dS$$

From Kelvin's circulation theorem, $\Gamma = 0$ at any time t. Then above expression reduced to

$$\int_S \nabla \times v.dS = 0 \quad \text{-----(1)}$$

Since S is arbitrary, equation (1) shows that $\vec{\Omega} = \vec{\nabla} \times \vec{v} = curl(\vec{v}) = 0$ at any time t.

INTRODUCTION TO FLUID KINEMATICS

Thus, fundamental axiom of irrotational flow is that vorticity is zero, mathematically,

$$\vec{\Omega} = \vec{\nabla} \times \vec{v} = curl(\vec{v}) = 0$$

From the fundamentals of fluid mechanics, we know that, $\vec{v}$ is the velocity vector, the curl of which is the vorticity vector, $\vec{\Omega}$ is zero, then vector can be expressed as the gradient of a scalar function ϕ, called the velocity potential function. Mathematically,

$$\vec{\nabla} \times \vec{v} = 0 \cong \vec{\nabla} \times (\vec{\nabla}\phi)$$

$$\Rightarrow \vec{v} = \vec{\nabla}\phi$$

Therefore, irrotational regions of flow field are commonly known as region of potential flow.

Recall the continuity equation for a steady, incompressible, planar, irrotational region of the flow field in the XY plane in cartesian coordinates,

$$\vec{\nabla} \cdot \vec{v} = 0 \Rightarrow \vec{\nabla} \cdot \vec{\nabla}\phi = 0 \Rightarrow |\vec{\nabla}|^2 \phi = 0$$

$$\Rightarrow \nabla^2 \phi = 0$$

Which is known as Laplace equation and scalar operator, ∇^2 Laplacian operator defined as $\vec{\nabla} \cdot \vec{\nabla}$, In cartesian coordinates,

$$u = \frac{\partial \phi}{\partial x} \quad v = \frac{\partial \phi}{\partial y}$$

$$\Rightarrow \nabla^2 \phi = 0 \Rightarrow \frac{\partial^2 \phi}{\partial x^2} + \frac{\partial^2 \phi}{\partial y^2} = 0$$

Laplace equation is valid for three-dimensional flow fields whose cartesian coordinates are

$$u = \frac{\partial \phi}{\partial x} \quad v = \frac{\partial \phi}{\partial y} \quad w = \frac{\partial \phi}{\partial z}$$

$$\because \nabla^2 \phi = 0 \Rightarrow \frac{\partial^2 \phi}{\partial x^2} + \frac{\partial^2 \phi}{\partial y^2} + \frac{\partial^2 \phi}{\partial z^2} = 0$$

In cylindrical coordinates,

$$v_r = \frac{\partial \phi}{\partial r} \quad v_\theta = \frac{1}{r}\frac{\partial \phi}{\partial \theta} \quad v_z = \frac{\partial \phi}{\partial z}$$

$$\because \nabla^2 \phi = 0 \Rightarrow \frac{1}{r}\frac{\partial}{\partial r}\left(r\frac{\partial \phi}{\partial r}\right) + \frac{1}{r^2}\frac{\partial^2 \phi}{\partial \theta^2} + \frac{\partial^2 \phi}{\partial z^2} = 0$$

Clearly, it's observed that three unknown fluid motion's components, $u, v,$ and w or $v_r, v_\theta,$ and v_z, depending on our choice of coordinate system substitute by one unknown variable, ϕ. Solving the two equations in terms of scalar variable, ϕ, and pressure, P, we obtain the value for ϕ which is used to determine the value of three components of the fluid motion.

The beauty of laplace equation is that its adaptability into various fields of engineering and physics as well. However, solution of laplace equation are depends on the boundary conditions of the flow field. As the laplace equation derived from the conservation of mass but mass (or density or volume) is not included in the equation. Once we solve the laplace equation for ϕ, with a given set of boundary conditions of flow field irrespective of the fluid properties, we can determine the fluid motion without solving Navier-Stokes equation. Since the laplace equation does not depends on the density or viscosity, so its solution is valid for any inviscid fluid consist of rotational regions and irrotational regions of the flow. The laplace equation is valid for an unsteady flow as any instant of time is not included in the continuity equation.

FLOW THROUGH CARTESIAN PLANES

For inviscid fluid flow in $XY-$ plane, the stream function ψ is defined as

INTRODUCTION TO FLUID KINEMATICS

$$u = \frac{\partial \psi}{\partial y} \qquad v = -\frac{\partial \psi}{\partial x} \qquad \text{-------(1)}$$

It must be keep in mind that stream function is defined in such a way that it valid for rotational as well as irrotational region of flow field. Let us consider only irrotational region of flow field, then vorticity is zero or we can say that $z-$ component of vorticity is zero. Mathematically,

$$\Omega_z = \frac{\partial v}{\partial x} - \frac{\partial u}{\partial y} = 0$$

$$\Rightarrow \frac{\partial}{\partial x}\left(-\frac{\partial \psi}{\partial x}\right) - \frac{\partial}{\partial y}\left(\frac{\partial \psi}{\partial y}\right) = 0$$

$$\Rightarrow -\left[\frac{\partial^2 \psi}{\partial x^2} + \frac{\partial^2 \psi}{\partial y^2}\right] = 0$$

$$\Rightarrow \nabla^2 \psi = 0 \qquad \text{-------(2)}$$

Thus, we can say that laplace equation is applicable, not only for velocity potential, ϕ, but also for stream function, ψ in steady, irrotational region of flow field. On solving respective laplace equations, curves of constant values of ϕ determine equipotential lines, and curves of constant values of ψ determine streamlines of a flow field. It has observed in irrotational region of flow, streamlines and equipotential lines are mutually orthogonal to each other as shown in figure. So, we refer that stream function, ψ and velocity potential function, ϕ are harmonic functions, and complementary to each other.

FLOW THROUGH AXISYMMETRIC AXES

The flow through axisymmetric axes is a unique case of two-dimensional flow which can be express in cylindrical coordinates. In this case, fluid flow is independent of rotational component of vorticity as rotational symmetry is defined about the $z-$ axis as shown in figure. In other words, dimension of the axisymmetric body and velocity field have no dependence on angle θ, such that $v_\theta = 0$. Also, the magnitudes of the fluid motion's components v_r and v_z remain unchanged irrespective of rotationality.

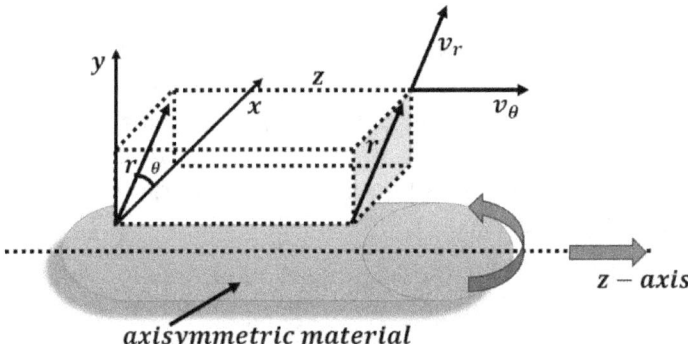

axisymmetric material

In the case of flow through axisymmetric axes, velocity potential ϕ satisfies the laplace equation,

$$\frac{1}{r}\frac{\partial}{\partial r}\left(r\frac{\partial \phi}{\partial r}\right) + \frac{\partial^2 \phi}{\partial z^2} = 0$$

For inviscid fluid flow in $XY-$ plane, the stream function ψ is defined as

$$v_r = -\frac{1}{r}\frac{\partial \psi}{\partial z} \quad \text{and} \quad v_z = \frac{1}{r}\frac{\partial \psi}{\partial r}$$

It must be keep in mind that stream function is defined in such a way that it valid for rotational as well as irrotational region of flow field. Since, $\theta-$ component of vorticity is zero or negligibly small due to which the fluid motion lies in the $rz-$ planes for axisymmetric irrotational regions of flow field. Thus, it is considered as two-dimensional flow always. Mathematically,

$$v_\theta = \frac{\partial v_r}{\partial z} - \frac{\partial v_z}{\partial r} = 0$$

$$\Rightarrow \frac{\partial}{\partial z}\left(-\frac{1}{r}\frac{\partial \psi}{\partial z}\right) - \frac{\partial}{\partial r}\left(\frac{1}{r}\frac{\partial \psi}{\partial r}\right) = 0$$

Since r is not function of z. Then above expression reduce to

$$r\frac{\partial}{\partial r}\left(\frac{1}{r}\frac{\partial \psi}{\partial r}\right) + \frac{\partial^2 \psi}{\partial z^2} = 0$$

We should keep in mind that above expression is not laplace equation for stream function, ψ. You cannot apply the laplace equation for the stream function for the axisymmetric irrotational regions of flow field. However, above expression is a linear partial differential equation which can be used for superposition of flows.

It has observed that the laplace equation is valid for velocity potential function, ϕ but not for the stream function, ψ for an axisymmetric irrotational regions of flow field. As a result, curves of constant ϕ and curves of constant ψ for an axisymmetric irrotational regions of a flow field are not mutually orthogonal.

CONSTITUTIVE EQUATION FOR IRROTATIONAL FLOW

Consider the Euler equation for an inviscid flow as

$$\rho\left[\frac{\partial \vec{v}}{\partial t} + (\vec{v}.\vec{\nabla})\vec{v}\right] = -\nabla P + \rho\vec{g} \quad \text{------(1)}$$

Assume that inviscid fluid flow is steady and incompressible, so equ (1) reduced to

INTRODUCTION TO FLUID KINEMATICS

$$(\vec{v}.\vec{\nabla})\vec{v} = -\frac{\nabla P}{\rho} + \vec{g}$$

$$\Rightarrow (\vec{v}.\vec{\nabla})\vec{v} = -\nabla\left(\frac{P}{\rho}\right) + \vec{g} \quad \text{------(2)}$$

Recall operations from vector algebra, advective term $(\vec{v}.\vec{\nabla})\vec{v}$ can be rewritten as

$$(\vec{v}.\vec{\nabla})\vec{v} = \vec{\nabla}.\left(\frac{v^2}{2}\right) - \vec{v}\times(\vec{\nabla}\times\vec{v})$$

$$\Rightarrow (\vec{v}.\vec{\nabla})\vec{v} = \vec{\nabla}.\left(\frac{v^2}{2}\right) - \vec{v}\times\vec{\Omega} \quad \left[\because \vec{\Omega} = \vec{\nabla}\times\vec{v}\right]$$

Next, Euler equation given by equ (2) reduce to

$$\vec{\nabla}.\left(\frac{v^2}{2}\right) - \vec{v}\times\vec{\Omega} = -\vec{\nabla}\left(\frac{P}{\rho}\right) + \vec{g} \quad \text{------(3)}$$

Assume that the acceleration due to gravity, $\vec{g}$ acts along the $z-$ direction only, we have

$$\vec{g} = -g\vec{k} = -g\vec{\nabla}z = \vec{\nabla}.(-gz) \quad \left[\because \vec{\nabla}z = \underbrace{\frac{\partial z}{\partial x}}_{0}\vec{i} + \underbrace{\frac{\partial z}{\partial y}}_{0}\vec{j} + \underbrace{\frac{\partial z}{\partial z}}_{1}\vec{k} = \vec{k}\right]$$

Next, using above relation, equ (3) reduce to

$$\vec{\nabla}.\left(\frac{v^2}{2}\right) - \vec{v}\times\vec{\Omega} = -\vec{\nabla}\left(\frac{P}{\rho}\right) + \vec{\nabla}.(-gz)$$

$$\Rightarrow \vec{\nabla}.\left(\frac{v^2}{2}\right) + \vec{\nabla}.\left(\frac{P}{\rho}\right) + \vec{\nabla}.(gz) = \vec{v}\times\vec{\Omega}$$

$$\Rightarrow \vec{\nabla}.\left(\frac{P}{\rho} + \frac{v^2}{2} + gz\right) = \vec{v}\times\vec{\Omega} \quad \text{------(4)}$$

In case of the irrotational region of fluid flow field where vorticity vector is almost zero, so that equ (4) can be rewritten as

$$\vec{\nabla}\cdot\left(\frac{P}{\rho}+\frac{v^2}{2}+gz\right)\equiv 0$$

As we know that if the gradient of scalar quantity is zero, then scalar quantity must be constant. So, we can derive the constitutive equation for irrotational regions of any fluid flow field,

$$\frac{P}{\rho}+\frac{v^2}{2}+gz=K\ (\text{constant})$$

The above equation refer to constitutive equation for irrotational region any fluid flow field because it holds everywhere throughout the regions of flow. We can find out the fluid motion by solving the laplace equation for velocity potential function, ϕ provided boundary conditions of the flow field is mentioned. Once the fluid motion is calculated, then we can determine pressure field by solutions of constitutive equation for irrotational region of flow field.

KINETIC ENERGY POSSES BY IRROTATIONAL FLOW

Kinetic energy possess by irrotational flow through open ended surface is the summation of the products of induced pressure and half the fluid motion. Thus, kinetic energy of a given mass of a fluid flow irrotationally in simply-connected region depends on the fluid motion along with boundaries of the region.

Consider an inviscid fluid of volume V flow irrotationally, at rest at infinity, and bounded internally by a surface S. Let ϕ_1, ϕ_2 are both single valued and continuously differentiable velocity potential functions and $\partial \phi_i / \partial r$ is the fluid motion in normal direction.

From green's theorem, we have

INTRODUCTION TO FLUID KINEMATICS

$$\int_V (\nabla\phi_1 . \nabla\phi_2) \, dV = -\int_S \phi_1 \frac{\partial \phi_2}{\partial r} \, dS - \int_V \phi_1 \nabla^2 \phi_2 \, dV \quad \text{------(1)}$$

Assume that $\phi_1 = \phi_2 = \phi$ such that $\nabla^2 \phi_2 = 0$, then equ (1) reduced to

$$\int_V \nabla\phi . \nabla\phi \, dV = -\int_S \phi \frac{\partial \phi}{\partial r} \, dS$$

$$\Rightarrow \int_V \left[\left(\frac{\partial \phi}{\partial x}\right)^2 + \left(\frac{\partial \phi}{\partial y}\right)^2 + \left(\frac{\partial \phi}{\partial z}\right)^2 \right] dV = -\int_S \phi \frac{\partial \phi}{\partial r} \, dS \quad \text{------(2)}$$

Let us consider v be the fluid motion and ρ be the density of the fluid flow, then equ (2) reduced to

$$\int_V v^2 \, dV = -\int_S \phi \frac{\partial \phi}{\partial r} \, dS$$

$$\Rightarrow \frac{1}{2}\rho \int_V v^2 \, dV = -\frac{1}{2}\rho \int_S \phi \frac{\partial \phi}{\partial r} \, dS$$

$$\Rightarrow K.E. = \frac{1}{2}\rho \int_V v^2 \, dV = -\frac{1}{2}\rho \int_S \phi \frac{\partial \phi}{\partial r} \, dS \quad \text{-----(3)}$$

Where $\rho\phi$ is the induced pressure that makes the fluid flows irrotationally from the rest, and $-\partial\phi/\partial r$ is the fluid motion acts in inward normal direction.

Deduction: Let $\partial\phi/\partial r = 0$ on the boundary of the surface. Then equ (3) reduced to

$$K.E. = \frac{1}{2}\rho \int_V v^2 \, dV = 0$$

$$\Rightarrow \int_V v^2 \, dV = 0$$

$$\Rightarrow v = 0 \quad \left[\because v^2 = +ve \text{ value} \right]$$

In other words, fluid motion is zero or negligible means that the inviscid fluid flow is at rest and there is no possibility of any cyclic irrotational movement of the fluid when it is confined by boundaries.

BOOSTER CAPSULE

Q 1. What is the difference between *vorticity* and *circulation*?

Answer: Vorticity and circulation are the two important parameters for the rotation in a fluid. Circulation, is a macroscopic measure of rotation for a definite space of a fluid. Also, vorticity, is a microscopic measure of rotation calculated at the fixed point in the space of a fluid. Moreover, circulation is considered as scalar quantity, and vorticity is a vector quantity specified to a fluid element.

Q 2. Differentiate between the vorticity and vortex?

Answer: Vorticity is the mathematical representation of vortex, while vortex is a physical phenomenon in nature with relation to the rotation in a fluid.

Q 3. Prove that the necessary and sufficient condition for the vortex lines are at right angles to the stream lines is

$$u, v, w = \mu\left(\frac{\partial \phi}{\partial x}, \frac{\partial \phi}{\partial y}, \frac{\partial \phi}{\partial z}\right),$$

Where μ, ϕ are functions of x, y, z, t

Solution. Streamlines are given by

$$\frac{dx}{u} = \frac{dy}{v} = \frac{dz}{w} \quad ------(1)$$

And vortex lines are given by

$$\frac{dx}{\Omega_x} = \frac{dy}{\Omega_y} = \frac{dz}{\Omega_z} \quad ------(2)$$

(1) and (2) will be at right angles, if

$$u\Omega_x + v\Omega_y + w\Omega_z = 0 -----(3)$$

But

$$\Omega_x = \frac{\partial w}{\partial y} - \frac{\partial v}{\partial z}$$

INTRODUCTION TO FLUID KINEMATICS

$$\Omega_y = \frac{\partial u}{\partial z} - \frac{\partial w}{\partial x} \quad \text{and}$$

$$\Omega_z = \frac{\partial v}{\partial x} - \frac{\partial u}{\partial y} \quad ----(4)$$

Using (4), (3) can be rewritten as

$$u\left(\frac{\partial w}{\partial y} - \frac{\partial v}{\partial z}\right) + v\left(\frac{\partial u}{\partial z} - \frac{\partial w}{\partial x}\right) + w\left(\frac{\partial v}{\partial x} - \frac{\partial u}{\partial y}\right) = 0$$

Which is the necessary and sufficient condition in order that $udx + vdy + wdz$ may be a perfect differential. So, we may write

$$udx + vdy + wdz = \mu\, d\phi = \mu\left(\frac{\partial \phi}{\partial x}dx + \frac{\partial \phi}{\partial y}dy + \frac{\partial \phi}{\partial z}dz\right)$$

$$\therefore\ u = \mu\left(\frac{\partial \phi}{\partial x}\right) \quad v = \mu\left(\frac{\partial \phi}{\partial y}\right) \quad w = \mu\left(\frac{\partial \phi}{\partial z}\right)$$

Q 4. Show that the components of velocity of an incompressible fluid u, v, w are solutions of Laplace's equation, given that the vorticity at every point is constant in magnitude and direction

Solution. Let $\Omega_x, \Omega_y, \Omega_z$ be the components of vorticity Ω so that

$$\Omega = \left(\Omega_x^2 + \Omega_y^2 + \Omega_z^2\right)^{1/2}$$

And the direction cosines of its direction are

$$\frac{\Omega_x}{\Omega},\ \frac{\Omega_y}{\Omega},\ \frac{\Omega_z}{\Omega}$$

Moreover,

$$\left.\begin{array}{l}\Omega_x = \dfrac{\partial w}{\partial y} - \dfrac{\partial v}{\partial z} \\ \Omega_y = \dfrac{\partial u}{\partial z} - \dfrac{\partial w}{\partial x} \\ \Omega_z = \dfrac{\partial v}{\partial x} - \dfrac{\partial u}{\partial y}\end{array}\right\} ----(1)$$

Differentiating second equation in (1) w.r.t. 'z' and third

INTRODUCTION TO FLUID KINEMATICS

equation in (1) w.r.t. 'y' and then subtracting, we get

$$\frac{\partial^2 u}{\partial z^2} - \frac{\partial^2 w}{\partial x \partial z} - \frac{\partial^2 v}{\partial y \partial x} + \frac{\partial^2 u}{\partial y^2} = 0$$

$$\Rightarrow \frac{\partial^2 u}{\partial z^2} + \frac{\partial^2 u}{\partial y^2} - \frac{\partial}{\partial x}\left(\frac{\partial v}{\partial y} + \frac{\partial w}{\partial z}\right) = 0 ----(2)$$

The equation of continuity is

$$\frac{\partial u}{\partial x} + \frac{\partial v}{\partial y} + \frac{\partial w}{\partial z} = 0$$

$$\frac{\partial v}{\partial y} + \frac{\partial w}{\partial z} = -\left(\frac{\partial u}{\partial x}\right) ---- (3)$$

Using (3), (2) reduces to

$$\frac{\partial^2 u}{\partial x^2} + \frac{\partial^2 u}{\partial y^2} + \frac{\partial^2 u}{\partial x^2} = 0$$

Showing that u satisfies Laplace's equation. Similarly, we can show that v and w also satisfy Laplace's equatio

HIGHER ORDER THINKING SKILL QUESTION

Q. Calculate the pressure field in a tornado.

Solution: Consider the flow through a tornado is two-dimensional represents in two distinct regions in cylindrical coordinates planes. The inner region $(0 < r < R)$ is rotational region of an inviscid flow. The outer region $(r > R)$ is an irrotational region of an inviscid flow as shown in figure.

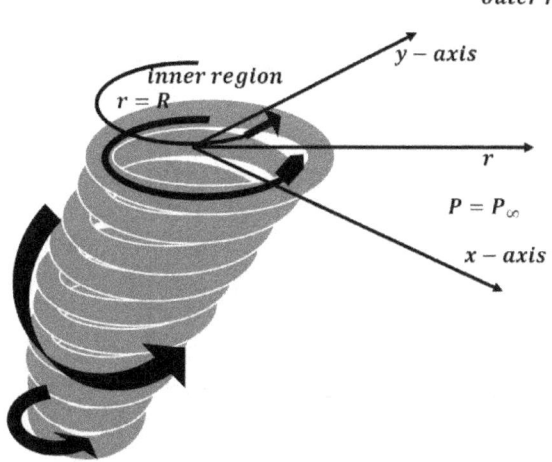

The components of the velocity $v = (v_r, v_\theta)$ of the flow is defined as

$$v_r = 0 \qquad v_\theta = \begin{cases} \omega r & 0 < r < R \\ \dfrac{\omega R^2}{r} & r > R \end{cases}$$

Where ω is the magnitude of the angular velocity in the inner region. The ambient pressure is equal to P_∞. Following assumptions are necessarily considered in order to determine the pressure field of the tornado.

1. The inviscid flow is steady and incompressible
2. Since radial distance expands and angular velocity

decreases with increment in height provided z, R and ω remain constants when we consider a particular horizontal slice.

3. The effects of gravitational force on any particular horizontal slice of tornado assumed to be insignificant.

Using assumption 1, we can apply Bernoulli's equation to the rotational region of inviscid flow as:

$$\frac{P}{\rho} + \frac{v^2}{2} + gz = C = \text{constant along streamlines}$$

$$\Rightarrow P = \rho C - \frac{1}{2}\rho v^2 - \rho gz \quad \text{-----(1)}$$

From the given condition, $v^2 = v_\theta^2 = \omega^2 r^2$ for any radial distance $0 < r < R$, then equation (1) reduced to

$$\Rightarrow P = \rho C - \frac{1}{2}\rho \omega^2 r^2 - \rho gz \quad \text{-----(2)}$$

At origin, $r = 0, z = 0$, the pressure is equal to P_i. and then equation (2) gives the value of C as

$$(2) \Rightarrow P_i = \rho C - 0 - 0$$
$$\Rightarrow P_i = \rho C$$

Put the value of ρC in the equation (2) and we obtain the expression for the pressure field as

$$P = P_i - \frac{1}{2}\rho \omega^2 r^2 - \rho gz \quad \text{------(3)}$$

Using assumption 3, we obtain the expression for pressure field in the inner region $(r < R)$ as

$$P = P_i - \frac{1}{2}\rho \omega^2 r^2 \quad \text{------(4)}$$

Since the outer region is a region of irrotational flow, the Bernoulli's equation is appropriate and Bernoulli constant is the same everywhere from $r = R$ $r \to \infty$.

$$(1) \Rightarrow P = \rho C - \frac{1}{2}\rho v^2 - \rho g z$$

$$\Rightarrow P = \rho C - \frac{1}{2}\rho v^2 \quad \text{------(5)} \quad [g = 0 \text{ on using assumption 3}]$$

We can determine the Bernoulli constant by applying boundary condition far from the tornado as

boundary conditions: $r \to \infty \Rightarrow v_\theta \to 0$ and $P \to P_\infty$.

$$\Rightarrow P_\infty = \rho C - 0 \Rightarrow P_\infty = \rho C$$

Put the value of ρC in the equation (5) and we obtain the expression for the pressure field in the outer region $(r > R)$ as

$$P = P_\infty - \frac{1}{2}\rho v^2 \quad \text{------(6)}$$

From the given condition, $v^2 = v_\theta^2 = \dfrac{\omega^2 R^4}{r^2}$ for any radial distance $r > R$, then equation (6) reduced to

$$P = P_\infty - \frac{1}{2}\rho \frac{\omega^2 R^4}{r^2} \quad \text{-----(7)}$$

At $r = R$, the interface between the inner and outer regions, the pressure must be continuous. Equating equation (4) and (7) at the interface region of inviscid flow, we get

$$P_{r=R} = P_i + \frac{1}{2}\rho \omega^2 R^2 = P_\infty - \frac{1}{2}\rho \frac{\omega^2 R^4}{R^2}$$

$$\Rightarrow P_i = P_\infty - \rho \omega^2 R^2 \quad \text{------(8)}$$

The equation (8) provides the value of lowest pressure in the middle of the tornado-eye of the cyclone. Put the value of P_i in equation (4) which enables us to rewrite equation (4) in terms of the ambient pressure P_∞,

$$P = P_\infty - \rho\omega^2 R^2 - \frac{1}{2}\rho\omega^2 r^2$$

$$\Rightarrow P = P_\infty - \rho\omega^2\left(R^2 + \frac{r^2}{2}\right) \quad \text{-----(9)}$$

From the equation (7) and equation (9) provides the pressure distribution where P as a function of radial location and it is given by

Inner region $(r < R)$: $\quad \dfrac{v_\theta}{\omega R} = \dfrac{r}{R} \text{ } | \text{ } \dfrac{P - P_\infty}{\rho\omega^2 R^2} = \dfrac{-1}{2}\left(\dfrac{r}{R}\right)^2 - 1$

Outer region $(r > R)$: $\quad \dfrac{v_\theta}{\omega R} = \dfrac{R}{r} \text{ } | \text{ } \dfrac{P - P_\infty}{\rho\omega^2 R^2} = -\dfrac{1}{2}\left(\dfrac{R}{r}\right)^2$

Clearly, in the inner region, P increases parabolically with radial distance even though the velocity increases. While in the outer region, pressure increases as the velocity decreases as it is directly obtained from the Bernoulli's equation.

www.ingramcontent.com/pod-product-compliance
Lightning Source LLC
Chambersburg PA
CBHW050056230526
45470CB00004B/1557